# Wildlife of Seychelles

John Bowler

WILDGuides

First published 2006 by **WILD***Guides* Ltd.

**WILD***Guides* Ltd.
Parr House
63 Hatch Lane
Old Basing
Hampshire
RG24 7EB

www.wildguides.co.uk

ISBN 1-903657-06-7



A catalogue record for this book is available from the British Library.

# Contents



## SPECIES ACCOUNTS

# Key to species types

**Birds**
*page 28*

**Land Mammals**
*page 86*

**Marine Mammals**
*page 92*

**Tortoises**
*page 98*

**Turtles**
*page 100*

**Geckos**
*page 104*

**Chameleon**
*page 110*

**Skinks**
*page 112*

**Snakes**
*page 116*

**Frogs**
*page 118*

**Caecilians**
*page 122*

**Butterflies**
*page 126*

**Moths**
*page 132*

**Dragonflies**
*page 134*

**Ant-lion**
*page 148*

**Beetles**
*page 148*

**Cockroach**
*page 150*

**Wasps and Bees**
*page 150*

**Millipede**
*page 152*

**Centipede**
*page 152*

**Whip Scorpion**
*page 152*

**Spiders**
*page 154*

**Crustaceans**
*page 156*

**Molluscs**
*page 164*

# Foreword

There is something captivating and compelling about islands, something that inspires the human spirit and captures the imagination. Their remoteness, their simplicity and their unique wonders add to their allure. The very existence of Seychelles is one of the wonders of the world. These are the only oceanic islands of granite on Earth. The oldest islands on the planet, separated from the continents for 65 million years. Nature has had time here to conduct its experiments in isolation and produce a unique wildlife. If you want sun, sea and sand then there are cheaper destinations but they will not have what Seychelles has to offer.

As survey after survey reaffirms, Seychelles has indeed got the best beaches in the world; but stray beyond the beach and there are more records. Seychelles has the world's largest seed, the world's smallest frog, a dozen species of birds found nowhere else, the world's only amphibians in an ocean setting, some of the most spectacular seabird colonies anywhere, around 75 endemic plants and the world's largest population of giant tortoises.

Seychelles is fortunate in that it has had a shorter period of human occupation than any other country on earth. There has simply been less time to mess things up. The usual story of indiscriminate destruction of fragile islands has not been played out here, partly from good luck and partly from good management. Damage has been done but it has not been terminal. On the management side, Seychelles was quick to realise that as a tourism destination, the one competitive advantage it possesses over the rest of the world is a relatively pristine environment. Forty-six percent of the landmass has been set aside as National Parks, Special Reserves or Protected Areas. This includes two World Heritage Sites, Aldabra the world's largest raised coral atoll and the Vallée de Mai, both managed by Seychelles Islands Foundation, established by Government decree shortly after independence. The Ministry of Environment and Natural Resources is one of the most important and most respected institutions of Seychelles. Environmental education is an integral part of the school curriculum and Seychelles Broadcasting Corporation devotes considerable time to nature programmes.

The non-Government sector has been active, too. The Nature Protection Trust of Seychelles (NPTS), founded in 1992, was the first environmental Non-Governmental Organisation (NGO) in Seychelles. Based on the island of Silhouette, the NPTS has run successful tortoise and terrapin conservation programmes and conducted surveys of the biodiversity of many other islands. Publications include the scientific journal *Phelsuma*, and the quarterly magazine *Birdwatch*, bringing together observations from around the islands.

The Island Conservation Society (ICS) founded in 2001 is another entirely local initiative. The ICS manages Aride Island Nature Reserve, the largest reserve of the granitics with the greatest number and variety of birds. The ICS has also commenced programmes in the relatively neglected outer islands of Seychelles including Cosmoledo, which had become the most important unprotected site in Seychelles. NGOs working hand in hand with the Government have achieved many victories, so that, for example, most of the endemic birds have increased in numbers, turtles which were once heavily exploited now receive protection, and respect for wildlife in Seychelles has become a way of life.

But one thing has been missing. It is all very well having a wonderful natural heritage but appreciation of this by as wide an audience as possible is enhanced by a well-illustrated popular guide book. This has been lacking until now. At last here is an identification guide to the wildlife of Seychelles, one of the most fascinating corners of the earth. With superb photographs including many species never before illustrated in any book and a clear and concise text, this is the culmination of years of research by the author and a celebration of Seychelles.

**Adrian Skerrett**
**Chairman of ICS & Treasurer of NPTS**

# Acknowledgements

A great many people have helped in the production of this book in a variety of ways and thanks are due to them all. The text was largely written when I was employed as warden of Aride Island Nature Reserve by The Royal Society for Nature Conservation. Sincere thanks are due to all members of the Aride Management and Scientific Committees, particularly James Cadbury, Adrian Skerrett, Libby Andrews and Tim Sands. Research materials used for the book were accumulated on Aride by many previous wardens, which made the task a little easier. Creole names for many of the species were given by Jimmy Pillay and José Hetimier, both Rangers on Aride Island. Others who helped in a variety of ways with research for the book within Seychelles include Edwin Palmer, John Nevill and Gerard Rocamora on Mahé, Serge Robert on Bird Island and Ron and Gill Gerlach on Silhouette.

I am most grateful to Adrian Skerrett for kindly writing the foreword and for assisting in locating some of the photographs. Additional expert help with the text was given by Bill and Chris Wain (Odonata), Justin Gerlach (invertebrates, reptiles, amphibians and mammals), Jeanne Mortimer (reptiles) and Adrian Skerrett (birds) to whom we extend our grateful thanks. I would also like to thank Rob Still and Andy Swash of **WILD***Guides* for their contribution, help and encouragement in the production of this book.

The photographic plates are one of the key features of this book and I would like to thank the following photographers who kindly allowed us to use their work: John C Abbott, Aurélien Audevard, Janet Bowler, John H Boyd III, Tim Cane, Kevin Carlson (Windrush), Colin Carver (Windrush), Roger & Liz Charlwood, Nigel Cleere, Trevor Codlin, Greg & Yvonne Dean, Stefano Fantini, Dagmar Fertl, Annika Forsten, Justin Gerlach, Amber Habib, Nic Hallam, Bruce Hallett, Jon Hornbuckle, Gordon Langsbury, Pat Lynch, Mark Lucas, Colin Macleod, Dave and Nancy Massie, Lori Mazzuca, Nigel McCall, Dick Newell, Tony Palliser, Gianni Pavan, Richard Revels (Windrush), Gerard Rocomora, Gehan de Silva, Adrian Skerrett, Rob Still, Andy & Gill Swash, Alan Tate, Chris & Bill Wain, Dylan Walker, Peter J Wilson, and Allan White.

I would particularly like to thank my wife Janet for her support and help during the production of the book.

John Bowler
March 2006

# Introduction

## About this book

The aim of this book is to enable anyone, be they visitor or resident, to identify the birds, land mammals, marine mammals, reptiles, amphibians, dragonflies or butterflies that they are most likely to observe in Seychelles. A less comprehensive section is devoted to some of the other larger and more conspicuous invertebrates that may also be encountered.

The comparative photographic plates and detailed text highlight the key differences between all similar species occurring in Seychelles, such that given suitable views, the subject can be identified with confidence. Differences between sexes have been illustrated where appropriate, together with distinctive immature plumages for the birds. The photographs for the plates were largely taken in Seychelles, and therefore illustrate the local races of wide-ranging species and the plumages that migratory birds are most likely to exhibit.

## How to use this book

The following sections provide information on the geography, climate and habitats of Seychelles *(pages 11–18)*, an insight into conservation in the region *(page 19)*, and a brief guide to some of the best places to observe wildlife *(pages 22–26)*.

The bulk of the book comprises stand-alone sections on the birds *(pages 28–85)*, mammals *(pages 86–95)*, reptiles and amphibians *(pages 97–123)* and invertebrates *(pages 124–165)*. Each section includes a brief introduction to the biological group, followed by a series of plates with accompanying text summarising the status of each of the species shown and highlighting the key identification features.

The final section of the book includes a glossary of the terms used *(page 166)*, suggestions for further reading *(page 167)*, useful addresses *(pages 168)*, a checklist summarising the distribution of the resident and regularly occurring species in Seychelles *(pages 168–173)*, a complete list of the birds species recorded *(pages 174–177)*, and a list of all the photographs and illustrations used in the book *(page 178-182)*.

Congo Rouge

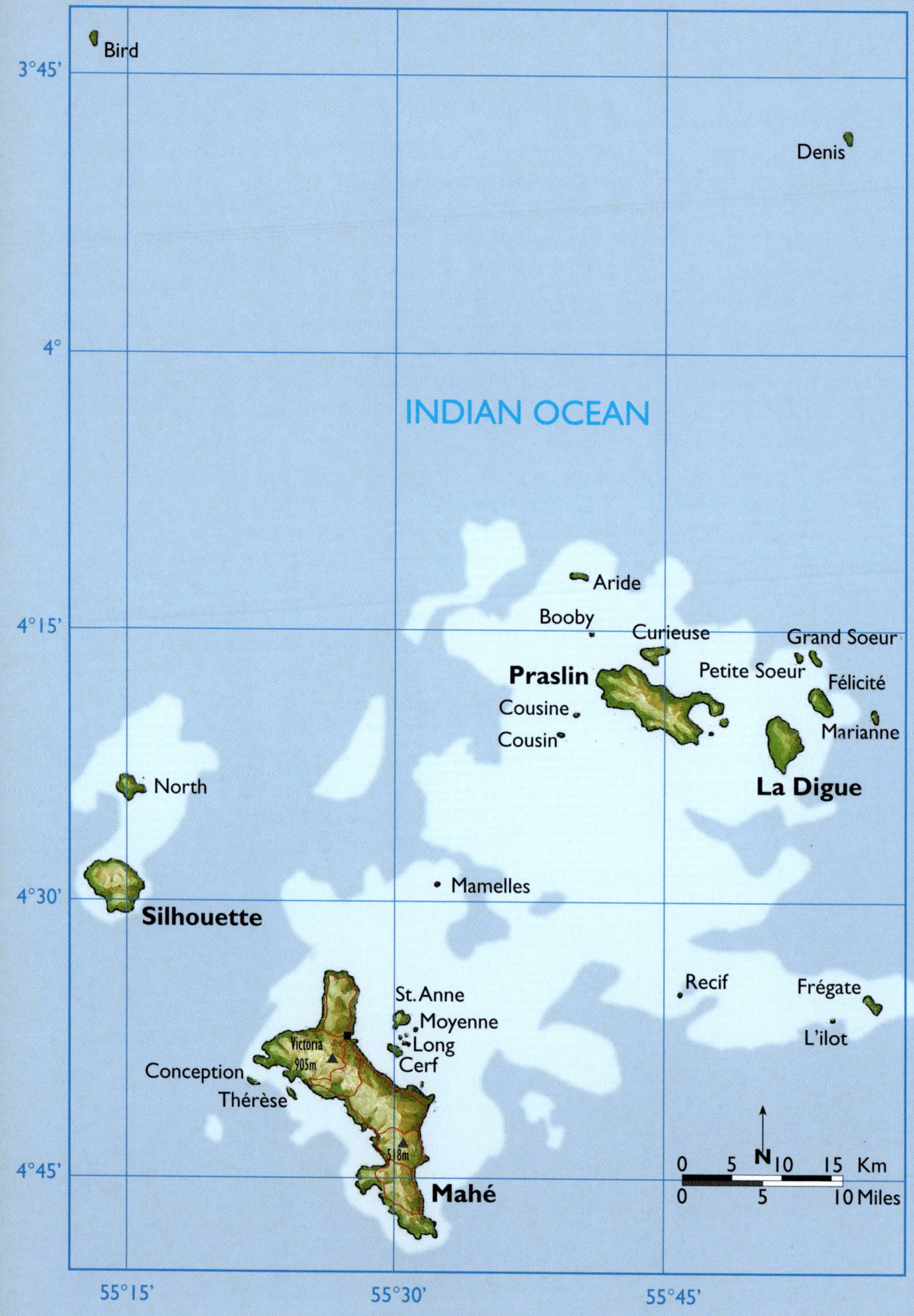

Bird
Denis
3°45'
4°
INDIAN OCEAN
Aride
Booby
Curieuse
Grand Soeur
Praslin
Petite Soeur
Félicité
Cousine
Cousin
Marianne
4°15'
North
La Digue
Mamelles
4°30'
Silhouette
Recif
Frégate
St. Anne
Moyenne
Long
Cerf
L'ilot
Victoria
905m
Conception
Thérèse
518m
4°45'
Mahé
N
0 5 10 15 Km
0 5 10 Miles
55°15'
55°30'
55°45'

# The Geography and Climate of Seychelles

## Geography

The Seychelles archipelago encompasses a large number of mostly small islands scattered across a vast swathe of the western Indian Ocean. The northern group of islands, centred on Mahé, is composed of granitic rock and comprises the only oceanic granitic islands in the world. They are thought to represent pieces of land that were left behind when the super-continent of Gondwana split up some 65 million years ago. The central granitic islands house the vast majority of the resident human population in Seychelles, and are the most accessible to visitors. Fortuitously, they are also the most interesting islands from a naturalist's perspective. With the notable exception of Aldabra, most of the 'outer' coralline islands are relatively depauperate in terms of the range of terrestrial habitats they support and the number of terrestrial species present. They are also either very difficult or very expensive for the average person to visit, and this book therefore deals only with the granitic islands and with the two adjacent flat coralline islands of Denis and Bird.

The largest island is Mahé, with a surface area of 150 km$^2$. This island hosts the capital Victoria and holds over 90% of the resident human population of around 82,000 people. It is a rugged, mountainous island, rising to 905 m (2,969 ft) at the summit of Morne Seychellois, the highest mountain in Seychelles. All the granitic islands have relatively steep rugged interiors, with flat areas of land known as 'plateau' restricted to the coastal fringes. On some islands, such as Praslin and La Digue, these plateau areas are quite extensive, whereas others, such as Félicité and Silhouette, have relatively small areas of plateau in relation to their size. The islands of Denis and Bird, which lie just north of the granitic islands, are of different origin, being coralline islands and are both flat and low-lying. All but the very smallest islands in the group are inhabited. Many of the smaller islands, such as Bird, Cousine, North and Frégate are run as relatively exclusive hotel resorts with a strong ecotourism element; whilst three, Aride, Cousin and Curieuse are operated as nature reserves which can be visited on day-trips. The size, highest point and human status of each of the main islands covered by this book are given in the table overleaf.

Cousin

| Island | Area (ha) | Altitude (m/ft) | Human Status |
|---|---|---|---|
| Mahé | 15,793 | 905 m/2,969 ft | High population |
| Praslin | 2,755 | 367 m/1,207 ft | High population |
| Silhouette | 1,995 | 740 m/2,428 ft | Low population |
| La Digue | 1,010 | 333 m/1,093 ft | High population |
| Curieuse | 285 | 172 m/364 ft | Nature reserve |
| Félicité | 270 | 231 m/758 ft | Resort |
| St. Anne | 220 | 250 m/820 ft | Resort |
| Frégate | 220 | 125 m/410 ft | Resort |
| North | 200 | 210 m/689 ft | Resort |
| Denis | 145 | 4 m/13 ft | Resort |
| Cerf | 105 | 108 m/354 ft | Low population |
| Bird | 100 | 4 m/13 ft | Resort |
| Marianne | 95 | 85 m/278 ft | Uninhabited |
| Grand Soeur | 85 | 111 m/364 ft | Low population |
| Thérèse | 75 | 160 m/525 ft | Uninhabited |
| Aride | 70 | 134 m/440 ft | Nature reserve |
| Conception | 60 | 120 m/397 ft | Uninhabited |
| Petite Soeur | 35 | 112 m/367 ft | Uninhabited |
| Cousin | 30 | 58 m/190 ft | Nature reserve |
| Cousine | 25 | 70 m/230 ft | Resort |
| Recif | 20 | *ca.* 30 m/100 ft | Uninhabited |
| Long | 20 | 85 m/279 ft | Low population |
| Moyenne | 10 | 60 m/197 ft | Low population |

## Climate

The climate of Seychelles is typically equatorial, being relatively hot and moist. However, conditions are moderated by the strong marine influence, such that temperatures are remarkably stable throughout the year. There are two distinct seasons: the 'northwest monsoon', which runs from November to March; and the 'southeast monsoon', which runs from May to September. Both seasons are named after the direction of the prevailing wind. The northwest monsoon is typically hotter and wetter than the southeast monsoon, and is characterised by calm, hot, sunny periods interspersed with blows from the northwest and heavy falls of rain. Wind strength is generally higher in the southeast monsoon, which is characterised by a constant breeze, rougher seas and cooler temperatures. Rainfall is normally lower during the southeast monsoon, but there is often more continuous cloud cover. The months of October and April fall between the two seasons and the winds may stop for a time and it can become very hot and humid, particularly in April, with temperatures rising to a maximum of 32·5°C. June is normally the coolest month, with a mean monthly maximum temperature of around 28°C, and temperatures at night dropping to a minimum of 23°C.

January is normally the wettest month, and daily deluges of over 60 mm are not uncommon. June and July are normally the driest months. The amount of rainfall differs markedly between the islands. The larger higher islands are generally wetter, and the highest peaks on Mahé and Silhouette are frequently draped in cloud, particularly during the northwest monsoon. These islands generate their own cloud patterns, and there is often a drier lee side, depending on wind direction. Smaller, lower-lying islands, such as Bird and Denis, which are more distant from the main islands, tend to be rather drier.

Average monthly temperatures (maximum and minimum) and rainfall are shown in the charts below.

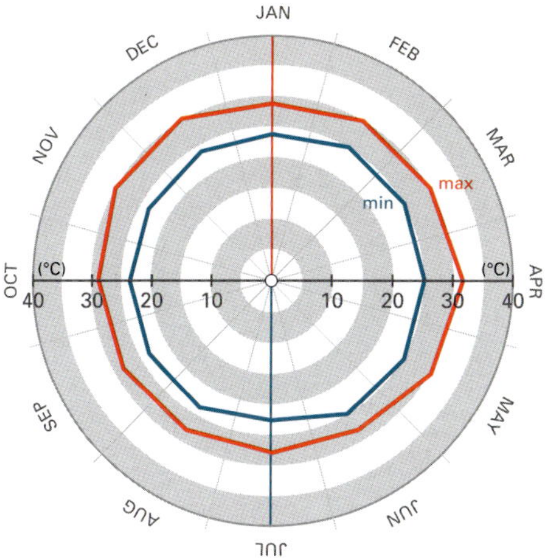


Average maximum and minimum monthly temperatures (°C)

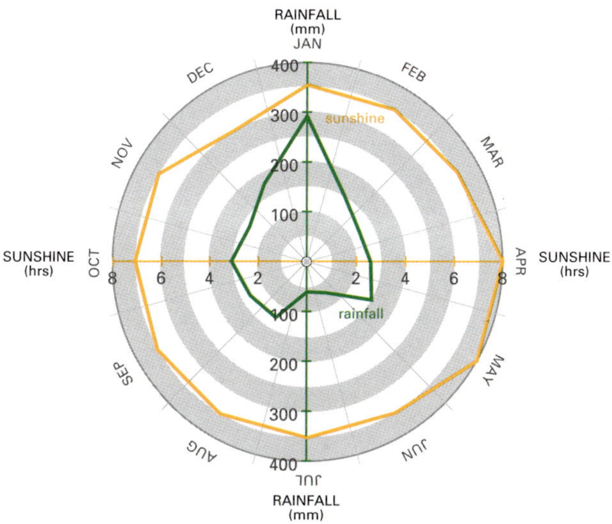


Average monthly rainfall (mm)

# The Habitats of Seychelles

There are eight distinct types of habitat within the northern Seychelles, including the open sea. All terrestrial habitats have, to a greater or lesser extent been altered by man, either through direct felling of the native vegetation and replacement by non-native species, or by the indirect effects of introduced plants and animals. Lowland habitats have, in general, been most altered by man, and little lowland forest remains, even in a semi-natural state. However, areas of habitat dominated by native plant species still persist at higher elevations, particularly on Mahé and Silhouette, as well as in the remarkable Vallée de Mai and Fond Ferdinand reserves on Praslin.

### The open sea

The seas around the northern Seychelles are shallow, mostly consisting of a raised shelf of 20–30m (65–100ft) depth, although there is a deeper channel (>40m (130ft) deep) between Mahé and Silhouette. The shallow Seychelles Bank extends north to include Bird and Denis Islands, but immediately north of these islands, the shelf drops steeply away to depths of up to 2,000m (6,500ft). The relatively shallow waters of the bank support a profusion of marine life including fringing coral reefs and locally important fisheries. Seabirds from the main colonies on Aride, Cousin, Cousine, Frégate and Bird are split between inshore feeders such as the noddies and Fairy Terns which feed on small fish close to the islands, and offshore feeders such as Sooty Terns and White-tailed Tropicbirds, which travel large distances to feed on the outer reaches of the bank.

### Rocky islets

There are numerous small rocky islets jutting up between the larger islands. Most are too small, too exposed or too low-lying to develop vegetation, but the larger islets such as Booby Island off Aride have developed salt-tolerant vegetation. Rats have generally not colonised or are unable to survive throughout the year, and, as a result, these islets often support surprising densities of nesting seabirds – evidence of the scourge that rats have been on other islands. Ground-nesting birds such as Brown Noddies and Bridled Terns dominate on the smaller islands, whereas vegetated islands also attract Lesser Noddies and Fairy Terns. Sooty Terns and Roseate Terns would once have nested, but excessive collection of eggs for human consumption has driven these species to the relative safety of the nature reserve islands.

## Inter-tidal areas

The large number of islands in Seychelles with their complex shorelines has resulted in the development of large areas of inter-tidal habitats. Most islands are surrounded by reef flats, and the shallow lagoons support a wealth of invertebrate life which in turn provides food for fish and inshore-feeding seabirds. The nature of the inter-tidal area itself is determined by the substrate, which in turn is influenced by wave action, tides, currents and the location of river outlets. Sheltered shores with a muddy substrate are generally most productive for invertebrates and provide attractive feeding grounds for migratory shorebirds and resident herons. Many of these areas have been lost to land reclamation and development, particularly on Mahé, but small patches remain. Several species of mangrove trees colonise the upper reaches of the mudflats, in turn adding nutrients to the mud and providing important nurseries for young fish, as well as providing habitat for mudskippers and fiddler crabs. White coral sand beaches dominate the more exposed inlets around the coasts, whilst the most exposed shores are steep and rocky, and have their own distinctive fauna including the Swift-footed Crab.

## Lowland forests

The flatter areas of land, or plateaux, near the coast contain the richest soils and, as a result, have been intensively settled and cultivated. The wooded areas that remain are much modified by man, but still retain a greater abundance of bird and animal life than the thinly-soiled hills. Semi-natural lowland forest is best seen on La Digue, where massive 'Takamaka' *Calophyllum inophyllum* and 'Bodanmyen' *Terminalia catappa* trees harbour the endemic Seychelles Paradise Flycatcher. Patches do, however, remain elsewhere on the larger islands, particularly just behind the beach-crests. The smaller seabird nature reserves of Aride and Cousin, have forests dominated by 'Mapou' *Pisonia grandis,* which afford good nest-sites for tree-nesting terns, and provide vital sanctuaries

for some of the most threatened endemic landbirds such as the Seychelles Warbler, Seychelles Fody and Seychelles Magpie-robin.

Most of the lowlands are dominated by introduced trees, including large numbers of fruit trees, and these support high densities of invertebrates, geckos, mammals such as the native fruit bat, and birds. The dominant lowland land-birds are Common Myna, Madagascar Fody and Barred Ground Dove, which are all introduced species, but the endemic Seychelles Sunbird also thrives here on the exotic flowering trees. Seychelles Bulbuls, Seychelles Blue Pigeons and Madagascar Turtle Doves also frequent the more heavily-wooded areas, and Seychelles Swiftlets hunt low over the plateaux at dusk on the large islands. On Mahé and Silhouette, Seychelles Kestrels frequent woodland edges in search of day geckos.

### Freshwater wetlands

This is the most severely threatened habitat type in Seychelles. Owing to the steep nature of the granitic islands, wetlands were mostly confined to the small areas of flatter land close to the coast, areas that have been settled intensively by man. Most wetlands were small, and many have been filled in to make way for housing and agriculture. The coralline islands of Bird and Denis have no surface water, as the coral bedrock is very porous. The largest remaining swamps are 'La Mare Soupap' on La Digue, Grand Barbe on Silhouette, and at the currently inaccessible Police Bay on Mahé. Smaller marshes remain scattered around most of the islands and all hold much of interest, including dragonflies, terrapins and caecilians along their margins, freshwater fish, and aquatic birds such as Moorhens and Yellow Bitterns. The large islands have streams and small rivers, which have a characteristic, although relatively depauperate, fauna including the two endemic mountain and stream damselflies.

## Coconut Plantations

Large areas of flatter land on the larger islands, and many of the smaller islands, were put over to the cultivation of coconuts for copra during the early twentieth century. As a result, large areas of lowland forest were lost. The copra trade has all but finished, but the coconuts remain. The plantations have a sparse fauna of introduced birds, but native geckos thrive in older plantations, together with widespread invertebrates such as the Coconut Beetle and Yellow Wasp.

On some reserve and private islands, indigenous trees have been allowed to recolonise old plantations and the coconuts have been removed to recreate native forests. This process has saved the formerly severely threatened Seychelles Warbler from extinction, and may hold the key to habitat improvement on a wider scale elsewhere.

## Mid-altitude forests

The mountains of the larger islands appear, at first sight, to be swathed in natural forest. Sadly, this is not the case, as the natural forest of the lower slopes (200–400 m (650–1,300 ft)) has mostly long since been removed and planted instead with forest crops of cinnamon and commercial timber trees, and on Mahé with tea bushes. The cinnamon has spread inexorably into adjacent forests, where it is joined by other alien species such as coco-plum and the enormous *Albizia* trees which dominate the mid-altitude forests on Mahé, and to a lesser extent on La Digue, Praslin and Silhouette. Surprisingly perhaps, these forests are rich in animal life. On Mahé, the Seychelles Scops-owl inhabits these forests, whilst the Seychelles White-eye clings on in isolated patches of secondary forests rich in invertebrates. Seychelles Bulbuls and Seychelles Blue Pigeons are common, Tiger Chameleons occur in path-side bushes, and rocky stream valleys hold numerous endemic frogs, as well as snakes and damselflies.

Native palm forests were once more widespread, but are now restricted to more rocky areas which were left uncultivated, such as La Reserve on Mahé, and to similar areas on Silhouette, Curieuse and Praslin. Fires have ravaged much of Praslin and Curieuse, and trees struggle to grow in the remaining thin soil. However, the creation of fire breaks – and perhaps luck – has saved two large patches of palm forest on Praslin, where the curious Coco-de-Mer palms support a unique fauna in the outstanding reserves of Vallée de Mai and Fond Ferdinand. Here, Seychelles Black Parrots and Seychelles Blue Pigeons nest amongst the swishing palms, large numbers of geckos, including the enigmatic Giant Bronze Gecko, patrol the dimly-lit trunks, and endemic snails and slugs inhabit the native vegetation.

### High altitude forests

High mountains above 400 m (1,300 ft) in altitude are confined to Mahé and Silhouette. Forests at these altitudes have been less affected by man, and small areas of primary forest exist on the steepest slopes and on the higher, more remote ridges. The highest peaks are frequently covered in cloud, and the native trees are draped in mosses and ferns under conditions of high humidity. Other than passing Seychelles Bulbuls and Seychelles Sunbirds, birds are generally scarce, but these areas are the last refuge of many endemic animals, including several species of frogs, skinks and butterflies, as well as a host of other invertebrates and plants. Many species are cryptic or nocturnal, but diligence will be rewarded on a visit to these last vestiges of primal Gondwana. Fortunately, much of the remaining habitat falls within the Morne Seychellois National Park on Mahé, or within areas of environmentally sensitive forestry management on Silhouette.

# Conservation

The vegetation cover of Seychelles has not been ravaged to the extent of many other Indian Ocean islands, such as Mauritius and Rodrigues, and it consequently retains a lush tropical appearance, together with many of its endemic animals. Non-native plant species are widespread, but many of the remaining native animals appear to have coped to a greater or lesser extent to this change in the flora. Conservation is fortunately high on the agenda of the Seychelles Government, which recognises the value of maintaining native biodiversity, and realised from an early date the important role that an intact environment plays, both to the lives and livelihood of the residents, and in attracting tourists to these magical islands. Ecotourism is likely to become increasingly important to the economy of Seychelles and will rely upon the maintenance of key habitats throughout the islands.

Large areas of the granitic islands and their associated inshore reefs have been designated as natural reserves by the Government and receive varying degrees of protection. Non-Governmental Organisations (NGOs) work closely with the Government to ensure appropriate management of private reserves and of threatened endemic species. The important seabird islands, for example of Aride, Cousin, Cousine and Bird are managed by the Island Conservation Society, Nature Seychelles, and private individuals, respectively, working together with the Government to conserve these internationally important sites.

It is pleasing to note that it now seems unlikely that any of the endemic bird species will become extinct in the short term. Indeed, the futures of the most threatened species – the Seychelles White-eye, Seychelles Magpie-robin, Seychelles Scops-owl, Seychelles Black Parrot and Seychelles Paradise Flycatcher – are looking increasingly secure in the medium-term. This is all due to the considerable collaborative efforts between the Government and NGOs, such as Nature Seychelles, the Marine Parks Authority, the Royal Society for Nature Conservation, the Island Conservation Society, the Seychelles Islands Foundation, and the Wildlife Clubs of Seychelles, together with privately-owned islands such as Bird, Conception, Cousine, Denis, Frégate and North, involving land acquisition, research, habitat management, captive management and education.

Providing supplementary food for Seychelles Magpie-robins

The restoration of native habitat on smaller islands, and the associated removal of invasive alien species, is providing an exciting new impetus to the conservation of threatened bird species. There is now a need for managers to extend this work to include other threatened endemic animals, including reptiles, amphibians and invertebrates. Pioneering work on the captive breeding of threatened reptile species is being conducted on Silhouette by the Nature Protection Trust of Seychelles.

Ringing White-tailed Tropicbirds

The islands face increasing pressure from development. Land for housing and industry is in short supply, whilst the building of further hotels promises the chance of further revenue. Land reclamation has eased the pressure, but this has been at the expense of inshore reef and mudflat habitats. The challenge facing Seychelles is to minimise the impact of future development in order to retain and enhance the existing native biodiversity. This will be for the benefit and enyoyment of residents and tourists alike, though most importantly to conserve its own intrinsic value.

A day visit to Aride

## Island Conservation Society

**www.islandconservation.net & www.arideisland.net**

The Island Conservation Society (ICS) was registered as an NGO in Seychelles in 2001. Its goal, as set out in the Articles of Association, is that: "The Society promotes the conservation and restoration of island ecosystems, sustainable development of islands, and awareness of their vulnerability and vital importance to the planet's biodiversity".

The ICS has managed Aride Island Nature Reserve since October 2003, under a lease from the Royal Society of Wildlife Trusts of UK. Aride is the largest nature reserve of the granitic islands and supports the highest number of seabird species (including two that do not breed on any other island in the area) plus five endemic land birds. It also has at least one (possibly two) endemic plant species and one of the highest densities of lizards on Earth.

The ICS has also been mandated to advise and manage environmental and conservation issues in the outer islands of Seychelles under a Memorandum of Understanding with the Government parastatal, Islands Development Company. Programmes include biodiversity studies of all the islands. On Cosmoledo, research and monitoring programmes have been initiated to determine accurately the identity and abundance of the main biota, to eradicate introduced predators and plants, and to reintroduce species which are extinct on the island but still present on Aldabra.

The ICS has played a leading role in the Seychelles White-eye recovery Programme and also works closely with like-minded partners in the islands including, most notably, the Nature Protection Trust of Seychelles (based on Silhouette Island) and the Seychelles Islands Foundation (managers of the World Heritage Sites of Vallée de Mai and Aldabra).

## Nature Protection Trust of Seychelles

Nature Protection Trust of Seychelles

**www.members.aol.com/jstgerlach**

The Nature Protection Trust of Seychelles (NPTS) was established in 1992 and registered under Seychelles law as a non-profit-making, Non-Governmental Organisation. The NPTS is dedicated to furthering the conservation of nature in the Seychelles islands, through informed scientific research to conserve species by conserving their habitats. Fulfilling its aim of promoting education, study and research in all fields of natural history the NPTS publishes two journals. The annual scientific journal, Phelsuma, deals with all aspects of nature within the entire western Indian Ocean and includes reports on all NPTS projects. Birdwatch is a quarterly publication containing natural history news and anecdotes from most of the island reserves. NPTS conservation projects are based on informed scientific research which aims to protect species by protecting their habitats. The NPTS is a member of the World Conservation Union (IUCN) and is represented on the Species Survival Commission of IUCN.

CONSERVATION PROJECTS INITIATED AND MANAGED BY THE NPTS INCLUDE:

**Seychelles Giant Tortoise Conservation Project** – captive breeding and reintroduction of two giant tortoise species though to have been extinct since 1840 until their rediscovery in 1996.

**Seychelles Terrapin Conservation Project** – captive breeding and ecological research on two Critically Endangered terrapin species.

**Indian Ocean Biodiversity Assessment 2000–2005** – assessment of the taxonomy and current status of all animal, plant and fungal species on all 115 of the Seychelles islands.

**Silhouette Conservation Project** – conservation management of the third largest of the Seychelles islands and the most important biodiversity hotspot in the islands.

# Islands Apart – a guide to wildlife watching

Wildlife is all around you in Seychelles, even at times in your hotel room or home. Every day spent in the islands will bring new experiences. The lowlands of the main islands are home to a range of widespread animals that can survive in man-made and semi-natural habitats. Introduced birds such as the Madagascar Fody, Common Myna and Barred Ground-dove dominate, but are joined by wide-ranging native invertebrates such as the African Grass Blue butterfly and Wandering Glider dragonfly, whilst endemic and brightly-coloured day geckos lurk in the shrubberies. The number of species unique to Seychelles, the island endemics, is high and each island has its own specialities. Many of the larger granitic islands have been separated by the sea for thousands of years and as a result new species have evolved. Other species were formerly more widespread but through changes brought about by man are now restricted to pockets of favoured habitat on different islands. In order to experience the full range of habitats and species it is necessary to travel widely throughout the islands. Fortunately, with planes and ferry services, such travel is now relatively easy and will enable you to encounter many of the species in this guide.

### MAHE

Although the main island of Mahé has the largest human population, it still retains some very important wildlife habitats. Lowland forests have all but gone, but the mountainous interior is still well forested. Much of the forestry is of non-native species, although even here much endemic wildlife still survives. This includes the Seychelles Scops Owl and a few Seychelles White-eyes, and the forests resound to the high-pitched calls of Gardiner's Frog. At higher altitudes, native montane forest persists to enchant with its endemic plants, invertebrates, frogs and elusive caecilians. A walk in the high mossy forests of the Morne Seychellois National Park is a must for anyone based on Mahé and is an antidote to the hustle and bustle in the heat of downtown Victoria. The coastal strip still retains some very valuable marshes and mangroves. A visit to the Inter-island Quay in Victoria will reveal a wide range of wading birds and terns, with fiddler crabs busy on the mudflats.

### PRASLIN

Praslin is easily reached by plane or ferry from Mahé and holds a special charm of its own. The beaches and coastal villages are generally quieter and more laid-back, whilst the lack of high mountains results in a drier climate. Fires have been a big problem on the island, sweeping through the dry forests with depressing regularity over the years. However, two pockets of endemic palm forest have survived: Fond Ferdinand and the world-famous Vallée de Mai. Remaining coastal marshes and pools still harbour a few terrapins and Yellow Bitterns, together with myriad dragonflies and crabs. Neighbouring Curieuse is worth a visit in its own right to see the captive Aldabran Giant Tortoises and to explore the mangrove trail.

## Vallée de Mai – The Last Eden

Stepping into the Vallée de Mai is like walking back in time. Preserved against the odds, somehow escaping the threats of fires and logging, this unique palm forest is the closest we can come to experiencing life on the ancient landmass of Gondwana. It is no wonder that General Gordon upon entering the valley for the first time in 1881, thought he had stumbled across the Garden of Eden. Protected as a National Park by the Seychelles Government since 1945, and now managed by the Seychelles Islands Foundation, a visit to this enchanting place is a must for all nature lovers.

Entering the closed forest, you immediately experience the cool, humid, shady conditions and can often hear nothing but the gentle swishing and clacking of the fronds of the gigantic Coco-de-Mer trees in the breeze. You may see very little at first, but slow, careful observation will reveal the hidden gems of this ancient place. The endemic palms harbour a range of animals found nowhere else in the world, including the White Slug and Coco-de-Mer Snail, whilst fragile mountain damsels flitter through the vegetation at the edge of mountain streams. Three species of bronze gecko inhabit the giant palms, including the enigmatic Giant Bronze Gecko whose eyes glow amber as it sips silently on water trickling over the Coco-de-Mer nuts. More easily seen, although harder to separate are the Bronze and Dwarf Bronze Geckos as they rest on the trunks, plus the more colourful day geckos.

Bird song is often absent at midday but in the cool of the early morning and at dusk, the valley echoes to the wild cackling of Seychelles Bulbuls, whilst Seychelles Sunbirds flit through the crowns in search of nectar. Seychelles Black Parrots give their presence away with their oddly high-pitched whistles. They can be elusive here, and are perhaps most easily seen high up in fruit trees around the car park at the entrance to the reserve, where they are often joined by Seychelles Blue Pigeons. The viewing shelters overlooking the valley are worth visiting if only to see how small and fragile this area is (just 18ha!), but also offer the chance of seeing the parrots and pigeons flying above the canopy, together with Seychelles Swiftlets, Seychelles Fruit Bats and, with luck, a Seychelles Kestrel.

## LA DIGUE

La Digue is reached from Praslin by regular ferry from Baie St. Anne. The ferry is sometimes a rustic sailing boat and is a foretaste of the even slower pace of life on the island itself, where the bicycle remains the main mode of travel. Seychelles Paradise Flycatchers are often not too hard to spot as they flit through the tall trees of gardens and woodlots on the plateau, and this is the best place in the islands to look for Common Waxbills. The marshes of La Mare Soupap are also a good place to look for terrapins, caecilians, Yellow Bitterns and a range of dragonflies.

## SILHOUETTE

Silhouette is well off the usual beaten track but is an essential place to visit for keen naturalists. The island hosts the headquarters of the Nature Protection Trust of Seychelles and its well-kept breeding pens house the only captive breeding populations of Seychelles Giant and Arnold's Tortoises, as well as terrapins. Walks into the rugged forested interior will reveal a wealth of endemic plant and animal life, including strong populations of caecilians and Brauer's Skinks. Endemic birds include some easily seen Seychelles Kestrels, whilst the Dauban marsh is a good spot for herons and egrets, including the recently colonising Black-crowned Night-heron, together with Giant Mangrove Crabs and various dragonflies.

## BIRD AND DENIS

Rather different again are the coralline islands of Bird and Denis. Bird supports a huge Sooty Tern colony, as well as some of the tamest wading birds in the world which wander about on the short grass between the chalets. Both islands often play host to rare and unusual migrants, particularly at the start of the northwest monsoon, because of their position at the northern edge of the archipelago. Bird harbours a large and increasing nesting population of Hawksbill and Green Turtles, as well as the only known population of Common House Geckos in the islands. Denis, on the other hand, has more extensive patches of native coastal woodland and has been used for translocations of endangered land-birds, including Seychelles Warbler and Seychelles Fody in 2004.

**Bird Island**

**Seabird cities**

No trip to Seychelles would be complete without a visit to one or more of the smaller islands with their bustling seabird colonies. From Praslin is it easy to catch one of the day-trip boats to Aride and Cousin. These two very special islands have long been run as nature reserves, to protect their breeding colonies of seabirds and to keep them free of alien species such as rats, which could have a devastating impact on the ground-nesting species. Of the two, Aride harbours the largest and most diverse range of seabirds, with over a million birds of some 10 species, including Roseate Tern, Sooty Tern and Red-tailed Tropicbird, which are absent from Cousin. The seabird breeding season peaks during the southeast monsoon, but landing on Aride cannot be guaranteed in this season because of its exposed south-facing beach. However, a visit in from May to July is an unforgettable experience and one that is well worth the effort. A visit to both islands is recommended at all times, as many seabirds are present year-round and the islands also host impressive densities of Wright's and Seychelles Skinks, Bronze Geckos and other endemic wildlife. Extensive habitat management has been undertaken on both islands to revert former coconut plantations to native coastal forest. As a result, they now act as refuges for some of the rarest endemic birds including Seychelles Magpie-robin, Seychelles Warbler and Seychelles Fody, together with Seychelles Sunbird, Seychelles Blue Pigeon and small numbers of turtle doves showing features of the original Seychelles red-headed form.

A Sooty Tern colony at dusk

As well as its birds and mammals, Seychelles is home to a huge range of fascinating plants, such as this Pitcher Plant

# Introduction to the Species Accounts

*The plates*

Species are numbered sequentially on the plates and the following annotations are used:

- 1 Adult in breeding plumage (species 1)
- 1n Adult in non-breeding plumage
- 1w Adult in winter plumage
- 1m Male
- 1f Female
- 1i Immature
- 1j Juvenile
- 1c Chick

*Species name*

For each species described, the accepted English name is followed by its scientific name and the common Seychellois name(s).

As explained on *page 125*, a number of the invertebrate species in Seychelles do not have a common English name. To aid the reader, a descriptive name has been given to these species and in such cases the 'English name' is annotated with a '+' symbol.

*Endemic species and subspecies*

The names of species that are endemic to Seychelles are preceded with an E and those which are represented by endemic subspecies with an e.

*Introduced, immigrant and accidental species*

Species that have been introduced to Seychelles are preceded with an I.

Species that occur regularly as non-breeding migrants are preceded with an m and species that are rare visitors with a V.

*Measurements*

The average total length is shown for all species. The measurements are given in centimetres or millimetres and inches (metres and feet in the case of cetaceans and the giant tortoises and turtles). For measurements of less than 30 cm, the equivalent length in inches is given to one decimal point). Measurements of the wingspan are also given for species with distinctive flight actions or silhouettes, where this is useful in identification. Details of weights are given where appropriate.

*Distribution and conservation status*

Text in the yellow tinted box summarises the status of the species in Seychelles, and includes notes on distribution and habitat preferences, and an indication of abundance.

The conservation status for species that are threatened with extinction is shown in RED.

Standard criteria for categorising the degree of threat that species face in the world have been adopted by The International Union for the Conservation of Nature and Natural Resources (IUCN). These categories are defined in the table on *page 29*.

*Species text*

**IDENTIFICATION/DESCRIPTION:** A summary of the key identification features and a brief description of the most noticeable colours and patterns of the species, highlighting differences between males, females and immatures where these are noticeable. Where appropriate, the text is subdivided into **ADULT**, **JUV (= JUVENILE)** or **IMM (=IMMATURE)**.

**VOICE:** A brief description of the usual calls or songs, where this is helpful for identification.

**BEHAVIOUR:** A description of the typical activities of the species.

**BREEDING:** Summary information on the life-cycle of the species, including breeding season(s), and, where appropriate, a description of nest sites and eggs, the duration of incubation and other aspects of breeding, such as courtship behaviour.

## BIRDS

A total of 205 bird species had been recorded in the northern (granitic) Seychelles by the end of 2005. The vast majority of these are rare migrant species (136 species), which occur less than annually, and non-breeding migrants (34 species), which occur at least annually. There are 35 extant breeding species of which 11 are endemic to the islands, occurring nowhere else in the world. One formerly resident introduced species, Grey-headed Lovebird *Agapornis canus*, which became extinct around 1977, is not included in this analysis.

A further three species (Green-backed Heron, Cattle Egret and Seychelles Black Parrot) are represented by endemic subspecies. The taxonomy of the Black Parrot of Praslin and Curieuse is currently being debated and it may ultimately attain specific status. Two additional endemic subspecies became extinct in historical times: the Green Parakeet *Psittacula eupatria wardi* and the Chestnut-flanked White-eye *Zosterops mayottensis semiflava*, and are not included here. The Madagascar Turtle Dove is represented by two subspecies: the nominate form *Streptopelia picturata picturata* is believed to have been introduced from Madagascar and has genetically swamped the indigenous Seychelles Turtle Dove *S. p. rostrata*. The latter is now virtually extinct, although individuals showing characteristics of this form still occur on Aride, Bird, Cousin and Cousine Islands.

The bird species recorded in the northern Seychelles can be divided into eight 'types' based on the broad habitats in which they occur. The number of species in each of these 'types' and their status is summarised in the table below.

| Type | Species recorded | Breeding species | Regular migrants | Rare migrants | Endemic species | Endemic subspecies |
|---|---|---|---|---|---|---|
| Seabirds | 41 | 10 | 13 | 18 | 0 | 0 |
| Waterbirds | 31 | 6 | 1 | 24 | 0 | 2 |
| Shorebirds | 47 | 0 | 18 | 29 | 0 | 0 |
| Diurnal raptors | 14 | 1 | 0 | 13 | 1 | 0 |
| Nightbirds | 4 | 1 | 0 | 3 | 1 | 0 |
| Aerial feeders | 10 | 1 | 1 | 8 | 1 | 0 |
| Larger landbirds | 18 | 7 | 0 | 11 | 1 | 2 |
| Smaller landbirds | 40 | 9 | 1 | 30 | 7 | 0 |
| **Total** | 205 | 35 | 34 | 136 | 11 | 4 |

The endemic Seychelles Blue Pigeon

The conservation measures for birds and their habitats in Seychelles implemented by the Government and non-governmental organisations (NGOs) have concentrated on the endemic species and the seabird colonies. Thankfully, these measures have generally been successful and the outlook for most of the species is optimistic. However, eight of the Seychelles endemics, and one of the vagrant species to have occurred in Seychelles fall into the threatened categories adopted by The International Union for the Conservation of Nature and Natural Resources (the World Conservation Union) (IUCN). A summary of these threatened categories, and the species concerned, is given below (from BirdLife International (2000) *Threatened Birds of the World*, as subsequently updated). It should be noted that this listing excludes taxa that are currently identified as endemic subspecies, such as Seychelles Black Parrot, which if raised to specific status would fall into the Critically Endangered category.

| Category | Definition | Seychelles species concerned |
|---|---|---|
| CRITICALLY ENDANGERED | Facing an extremely high risk of extinction in the wild in the immediate future | Seychelles Paradise Flycatcher (E)<br>Seychelles White-eye (E) |
| ENDANGERED | Not critical but facing a very high risk of extinction in the wild in the immediate future | Seychelles Scops-owl (E)<br>Seychelles Magpie-robin (E) |
| VULNERABLE | Not endangered but facing a high risk of extinction in the wild in the medium-term future | Madagascar Pond-heron (V)<br>Seychelles Kestrel (E)<br>Seychelles Swiftlet (E)<br>Seychelles Warbler (E)<br>Seychelles Fody (E) |
| *Endemic species are annotated with* (E) *and vagrant species with* (V) | | |

## Rare migrant bird species

The isolated position of Seychelles causes the archipelago to act as a magnet for disoriented migrant birds. Although many species of shorebird undergo an annual migration to spend the northern winter in the benign climate of Seychelles, other species appear less regularly and in very small numbers, and are presumably lost and wandering individuals. These vagrants can include almost any species migrating between Europe and Africa, intra-African migrants or even species migrating from mainland Eurasia to southeast Asia.

Many 'lost' birds are immatures, which can turn up almost anywhere and at any season. However, most vagrants turn up during the peak migration period of October to January, and they are often most apparent on smaller offshore islands. Bird, Aride, and Frégate are well-known hot-spots for vagrants, but other less well watched strategically-placed islands include Denis and North. The 136 rare migrant species which have been recorded from the northern Seychelles Islands up to and including 2005 are listed on *pages 174–177*. The majority of these birds have been recorded less than five times, but there are a handful of species that have turned up almost annually in recent years. The six rare migrants most likely to be seen are included in the following species accounts.

## Audubon's Shearwater (1)

*Puffinus lherminieri* Riga

Length: 29 cm/11·4 in  Wingspan: 69 cm/27 in

A fairly common resident, found in coastal waters around all the islands. Though breeding is restricted to a handful of rat-free islands, the vast majority nest on Aride. However, odd pairs may still attempt to nest in the mountains of the larger islands.

**IDENTIFICATION:** The small size and white underparts separate this species from the much larger Wedge-tailed Shearwater. In flight, the straight, narrow wings appear fairly short. **Adult:** Entirely blackish-brown above, and largely white below, with a conspicuously broad dark terminal edge and tip to the underwing, and a much narrower dark leading edge. The tail is dark, quite short and appears rather square-ended. The bill is small, slim and all-dark. The feet are pink but are not normally visible in flight. **Juv:** Very similar to adult.

**VOICE:** Adults give distinctive cackling notes at their breeding sites, especially upon arrival at nightfall and upon departure just before dawn, consisting of a descending series of rapid *"kuk–kuk"* notes terminating in a loud plaintive *"koww!"*. They are largely silent at sea.

**BEHAVIOUR:** Flies very low over the surface of the sea with rapid wing-beats interspersed with short glides on straight wings. Birds are often found far out to sea during the day, but gather in the late afternoon in floating rafts off the breeding islands. They visit their nest burrows at night, and many climb trees at dawn to reach high take-off points before heading back out to sea.

**BREEDING:** Birds are present on the breeding islands throughout the year, with nesting attempts occurring in 'waves'. A single white egg is laid in a burrow.

## Wedge-tailed Shearwater (2)

*Puffinus pacificus* Fouke

Length: 43 cm/17 in  Wingspan: 100 cm/39 in

A fairly common resident found in coastal waters around all the islands, though breeding is restricted to a handful of rat-free islands. Odd pairs may still attempt to nest on mountains of larger islands.

**IDENTIFICATION:** The only large all-dark shearwater likely to be seen in Seychelles waters. Its large size and dark underparts separate it from the much smaller Audubon's Shearwater. There is a white-breasted form but this is extremely rare in Seychelles. In flight, the broad-based wings are long and held slightly forward from the body, typically slightly bent at the carpal joint. **Adult:** Can appear blackish, but in good light is dark brown with a variably paler panel in the upperwing. The bill is largely dark grey but can show a paler base of bluish or greenish-grey, or even pinkish. The tail is quite long and typically appears pointed. The feet are pinkish. **Juv:** Very similar to adult, but pale edgings to the wing coverts form two indistinct bars. The bill is normally all-dark.

**VOICE:** Adults emit an eerie wailing *"woooh"* at their breeding sites and juveniles give various squeals, some not unlike the cries of a human baby. It is largely silent at sea.

**BEHAVIOUR:** Glides low over the surface of the sea on straight wings with only occasional flaps, characteristically banking up higher in irregular arcs. Birds are often far out to sea during the day, but stream past headlands in the late afternoon to gather in floating rafts off the breeding islands. They visit their nest burrows at night.

**BREEDING:** A single white egg laid in a burrow in September–October at the end of the southeast monsoon, with juveniles fledging in February–March. Small numbers, possibly failed breeders, lay eggs in June–July.

1
1
1
2
2
2

## Red-tailed Tropicbird

1

*Phaethon rubricauda* Payanke Lake Rouz

Length: 46 cm/18 in Wingspan: 100 cm/39 in (78 cm/31 in including tail streamer)

A very rare resident breeding only on the northern cliffs of Aride. It is extremely rare in coastal waters elsewhere.

**IDENTIFICATION:** The large size, almost completely white plumage, red bill and thin red tail streamers separate this species from the smaller and much more common White-tailed Tropicbird. The red tail streamers are difficult to see except at close range, giving birds a short-tailed appearance in flight. The wings are long and broad, and held stiffly when gliding. **ADULT:** Almost entirely snowy-white plumage, tinged pink especially across the breast, with black confined to the lower tertials and small eye-patch. The feet are large and conspicuously black in flight. The bill is large and red. **JUV:** Not often seen, as breeding is rarely successful. They are similar to juvenile White-tailed Tropicbirds but are larger, with more heavily barred black upperparts and a distinctive dark bill.

**VOICE:** The deep honking call is vaguely goose-like, and carries easily through the cacophony of higher-pitched tern calls when adults display over the breeding cliffs.

**BEHAVIOUR:** Circles on thermals over the breeding cliffs, sometimes approaching very close to observers. It is most active at the cliffs around midday, often flying out some 400 m (1/4 mile) or so to rest and preen on the surface of the sea. A spectacular cartwheeling display flight is exhibited by pairs, often attracting the attentions of single birds and frigatebirds.

**BREEDING:** Birds start to display over the breeding cliffs in September, and a single large egg is normally laid late in the year, with chicks fledging in March. Birds may be seen in most months of the year, but are more reliable in October–February. Breeding success in recent years has been very low and the species maintains a precarious toehold in the northern Seychelles.

## White-tailed Tropicbird

2

*Phaethon lepturus* Payanke Lake Blan

Length: 39 cm/15 in Wingspan: 90 cm/35 in (75 cm/30 in including tail streamer)

A common and widespread resident of coastal waters and often seen over hills inland.
It breeds in large numbers on rat-free offshore islands, and, although less numerous, also nests in tall trees and inland cliffs on the main islands.

**IDENTIFICATION:** Smaller in size than the Red-tailed Tropicbird, with a less bulky appearance, slimmer and shorter wings and more rapid wing-beats. Adults also have a striking black and white pattern on the upperwing and long, thin white tail streamers. Birds are often seen soaring high over the main islands, their stiffly-held wings, very long tail and black upperwing patterning separating them from the much smaller Fairy Tern *(page 60)*. **ADULT:** Largely white plumage, with a distinctive black bar across the upperwing coverts, and black base to the wing tips. The very long white tail streamers have a black shaft, although this is difficult to see except at close range. A narrow black bar curves from the gape up and through the eye, and there are also black markings on the thigh feathers. The bill is quite large and yellow, and the feet are very dark grey. **JUV:** Similar to the adult, but lacks tail streamers and has extensive blackish barring on crown, nape, mantle and wing coverts. The bill is yellowish, normally with a darker tip.

**VOICE:** Harsh, quite high-pitched *"kek, kek"* calls are given at the nest site and in aerial display.

**BEHAVIOUR:** Glides high over the breeding islands, pairs displaying in flight. It is very tame and approachable at nest sites on the ground. Adults feed extensively at night, and chicks are fed mostly in the morning. Birds are regularly seen resting and preening on the surface of the sea.

**BREEDING:** Breeding occurs throughout the year, pairs being largely faithful to their partner and nest site. The single brick-red egg takes around six weeks to hatch. The chicks take a further twelve weeks to fledge, normally walking from the nest site to the sea to do so.

2
1
2c
2j
2

## m Red-footed Booby 1

*Sula sula* Fou Bet

Length: 71 cm/28 in Wingspan: 100 cm/39 in

An uncommon and local visitor to coastal waters, especially around the seabird islands of Bird and Aride, where up to 50 (Bird) and 6 (Aride) individuals may roost amongst the frigatebirds. It can occur in any month, but is most regular in October–December and again in May–August.

**IDENTIFICATION:** This species always appears slimmer and less robust than the other species of booby. There are several different colour forms, ranging from white to brown, but in Seychelles the white form predominates. The red feet and legs of adults are diagnostic. The white form is similar to adults of the much larger, though scarcer, Masked Booby, but the tail is white and the dark mask of the latter species is absent. Darker, or intermediate, form birds are never as dark brown as adult Brown Booby or juvenile Masked Booby and have more uniformly coloured wings and a whitish, not dark, tail. **ADULT:** The white form is largely white with bold black wing tips, a black patch on the underside of the carpal joints, black trailing edges to the wings, and a yellowish cast to the head and nape. Intermediate form birds are variably greyish or sandy-brown, generally lacking strong plumage contrasts in the wing, and typically having a white head and tail. The slim, pointed bill is pale greyish-blue, often with a pinkish base. The large feet are prominently red. **JUV:** Largely grey-brown with yellowish or pinkish-orange legs.

## m Masked Booby 2

*Sula dactylatra* Fou Zenero

Length: 86 cm/34 in Wingspan: 152 cm/60 in

A rare but probably annual visitor to coastal waters around offshore seabird islands. It can occur at any time of year, but especially in January/February, and August.

**IDENTIFICATION:** The largest Booby in Seychelles waters, always appearing very large and robust, and gliding on long, broad-based wings. Most birds seen are in adult plumage and are best separated from the similarly plumaged white form of Red-footed Booby, by their larger size, black (not white) tail and blackish facial mask. Juveniles are similar to Brown Booby but have an obvious white collar, a paler mantle and more complex underwing pattern. **ADULT:** Mostly white above and below, with black flight feathers, a black tail and a prominent blackish facial mask. The large, pointed bill is yellowish, and the legs and feet are blackish. **JUV:** The dark brown head is separated from the paler brown mantle by an obvious white collar. The wings and tail are brown above. The underparts are white, including the throat, and there is a complex pattern on the underwing, with a dark brown line running through the otherwise white underwing coverts, and a darker trailing edge.

## m Brown Booby 3

*Sula leucogaster* Fou Kapisen

Length: 69 cm/27 in Wingspan: 141 cm/56 in

A rare but annual visitor to coastal waters, most often seen around offshore seabird islands, particularly Aride. It can occur at any time of year, and individual birds may linger for several months.

**IDENTIFICATION:** Appears stronger and more robust than the Red-footed Booby, gliding on long, broad-based wings. The dark brown upperparts, head and chest separate this species from the rarely seen juveniles of the Masked Booby, which are much larger, have an obvious white collar and paler mantle and more complex underwing pattern. **ADULT:** Uniformly dark brown upperparts head and chest, with all dark tail, and strongly contrasting white belly and central underwing. The bill is slim and pointed and the facial cere and feet are variably yellowish or greenish. **JUV:** As adult, but the underwing and underparts are mottled brown and show less contrast with the dark upperparts.

White-tailed Tropicbird shown for comparison

## Great Frigatebird

*Fregata minor* Gran Fregat

Length: 92 cm/36 in Wingspan: 218 cm/86 in

A common non-breeding visitor to coastal waters, also often seen soaring high inland over the larger islands. Up to 4,500 birds roost on Aride, where it greatly outnumbers the Lesser Frigatebird; smaller numbers roost on Bird and Cousin. The nearest breeding colonies are on Aldabra. Birds are present all-year on Aride, with numbers peaking in October/November.

**IDENTIFICATION:** Frigatebirds are immediately recognisable by their huge size, effortless gliding flight on very long slightly bowed wings, long forked tails and generally blackish plumage. Specific identification can be difficult and often requires prolonged, close views in good light. The Great Frigatebird is larger than Lesser Frigatebird and never shows the pronounced white patches extending from the sides of the chest onto the underwing which is characteristic of that species. **Adult:** Males are entirely black except for a red gular pouch, which is inflated during display. Females are all black (including the underwing), except for a whitish breast, throat and chin. **Juv:** Largely black but the head and belly are whitish or tawny. There is a brownish band across breast of very young birds, but this disappears with age. **VOICE:** Silent at sea, but birds give curious whinnying calls when roosting. **BEHAVIOUR:** Soars high over the sea and land on long, bowed wings, feeding predominantly by taking fish from the surface of the sea with their hooked bill. Frigatebirds also parasitise breeding seabirds, such as tropicbirds, shearwaters and noddies, chasing their victims relentlessly, forcing them to regurgitate their crop contents. On occasion they may bite their victim's wing, rendering them flightless.

## Lesser Frigatebird

*Fregata ariel* Pti Fregat

Length: 76 cm/30 in Wingspan: 184 cm/72 in

A regular non-breeding visitor to coastal waters, but much less numerous than Great Frigatebird with only tens of birds present in the large frigatebird roost on Aride, and even smaller numbers elsewhere.

**IDENTIFICATION:** Noticeably smaller than Great Frigatebirds when the two species are seen together. In all plumages the Lesser Frigatebird shows a pronounced and distinctive thin white patch extending from the sides of the chest onto the underwing. **Adult:** Male as male Great Frigatebird, except for diagnostic white stripe running from sides of breast onto underwing. Female differs from female Great Frigatebird in having a black throat and chin, a white stripe extending onto the underwing and a distinctive white marking on the side of the neck when seen from above. **Juv:** Similar to Great Frigatebird, but always shows a distinctive white marking extending onto the underwing. The head of very young birds may be darker than in Great Frigatebird of the corresponding age, whilst older juveniles show more black on the belly. **VOICE:** As Great Frigatebird. **BEHAVIOUR:** As Great Frigatebird. They are often seen chasing seabirds off the south coast of Aride in May–July (when the proportion of Lesser Frigatebirds to Great Frigatebirds is highest).

1m
1j
1f
1f
2m
2f
1j
2j

e

## Green-backed (Striated) Heron

1

Length: 40 cm/16 in Wingspan: 68 cm/27 in

*Butorides striatus* Mannik Ledo Ver

A common and widespread resident, occurring on most islands. It is found along shorelines of all types as well as in marshy areas and along streams inland. Birds are of the endemic subspecies *B. s. degens*.

**IDENTIFICATION:** A small, compact heron, which is larger and always appears darker than the much scarcer Yellow Bittern. **ADULT:** Largely greyish-green above, greyer on the neck, sides of breast and the belly, often showing a rich brown wash either side of a whitish line that runs down the centre of the throat and breast. The black crown is neatly demarcated, contrasting with the pale chin and lores. The bill is black, relatively large and dagger-like, and the legs and feet are yellowish-green (though the legs are often darker down the front), but are brighter yellow, even orange, when breeding. **JUV:** The upperparts are dark brown with pale fringes to the green-tinged wing feathers. The underparts are whitish, but heavily streaked dark brown. The crown is blackish but less clearly defined than in adult **(1j)**. **VOICE:** A distinctive sharp *"keyow!"* call is often given when flushed or in flight. **BEHAVIOUR:** Typically creeps along open muddy shorelines, over rocks or along wetland edges in search of prey (mostly small fish, frogs and invertebrates), but readily takes flight. **BREEDING:** May breed all-year round, although there appears to be a peak at the end of the northwest monsoon (December–April).

## Yellow (Chinese) Bittern

2

Length: 36 cm/14 in Wingspan: 53 cm/21 in

*Ixobrychus sinensis* Makak Zonn

A rare and localised resident of freshwater marshes on the main islands, and a rare inter-island migrant elsewhere. It often skulks in dense vegetation and can be difficult to observe.

**IDENTIFICATION:** Smaller and always more tawny-buff coloured than the more numerous Green-backed Heron. In flight, the boldly contrasting black and buff wing pattern is distinctive. **ADULT:** The crown and short tail are black, and the belly and breast are buff with diffuse dark lines particularly down the centre (most prominent in females). The relatively large, dagger-like bill is pale yellow, with a pronounced red base in breeding males. The legs and feet are yellow. **JUV:** Darker brown above and generally more heavily streaked than adults, with rusty fringes to the dark brown wing coverts. The belly is paler buff, with darker brown streaks **(2j)**. **VOICE:** Males give a repeated call of single *"hoo"* notes, not unlike the distant barking of a small dog, at dawn and dusk during the breeding season. **BEHAVIOUR:** Typically creeps up the stems of aquatic vegetation and may 'freeze' for lengthy periods. Birds sometimes emerge into more open areas in the early morning and at dusk and are often seen flying during the day. **BREEDING:** Poorly known. Males call during the latter half of the northwest monsoon (December–April).

## Black-crowned Night-heron

3

Length 40 cm/16 in Wingspan: 68 cm/27 in

*Nycticorax nycticorax* Manik Lannwit

Formerly only a vagrant, it has increased in numbers seen since 1996 and there is evidence of breeding on Mahé, Silhouette and Cousin; birds have also been seen foraging nocturnally on the west side of Praslin.

**IDENTIFICATION:** A mid-sized, stocky heron, typically seen in a hunched posture with its neck tucked well into its shoulders. The thick blackish bill, longer legs and larger size separate juveniles from the superficially similar juvenile Green-backed Heron. **ADULT:** The black mantle separates adults from all other herons. A well-defined black crown contrasts with the whitish face, neck and underparts, and the wings are uniform pale grey. The relatively short and thick bill is blackish, the legs and feet are yellow, and the eye is red. **JUV:** Much browner than adults: the upperparts are dark brown but extensively streaked and spotted with buffy-white; the underparts are whitish but with heavy dark brown streaking. The bill is blackish **(3j)**. **VOICE:** The call is a low, hoarse *"kwark"*, often given in flight. **BEHAVIOUR:** It is largely nocturnal in habit, although birds may also feed during the day, usually at coastal marshes and pools.

1

1j

2

2j

3

3j

## e Cattle Egret 1

*Bubulcus ibis* Madanm Paton

Length: 50 cm/20 in Wingspan: 90 cm/35 in

A common resident on Mahé where it is widespread in the lowlands, a local visitor to northwest Praslin from Cousin, and a scarce inter-island migrant elsewhere. Birds are of the endemic subspecies *B. i. seychellarum*.

**IDENTIFICATION:** A medium-sized, slim egret – the only all-white heron likely to be seen in Seychelles. The longish, slim neck and all white wings distinguish it in flight from distant tropicbirds. **ADULT:** The plumage is almost entirely white, but it attains rusty buff plumes on the crown, breast and mantle when in breeding dress. The relatively thick bill is ivory-yellow and the legs and feet are normally dusky, although they, and the base of the bill, may become bright pink in breeding plumage. **JUV:** Much as non-breeding adult. **VOICE:** Generally silent, but birds may give a harsh *"kraaa"* call upon taking flight. **BEHAVIOUR:** Hunts in open areas, short grassy fields and on mudflats around Victoria. It also frequents rubbish dumps and the Victoria fish market, where individuals are remarkably tame. Large numbers fly from feeding areas in the evening to congregate in favoured roosting trees. **BREEDING:** Breeding occurs mostly during the southeast monsoon (April–October), with notable colonies in and around the reclaim area of eastern Mahé.

## Grey Heron 2

*Ardea cinerea* Floranten Sann

Length: 95 cm/37 in Wingspan: 165 cm/65 in

A fairly common resident on Mahé where it is widespread in coastal areas, but a scarce inter-island migrant elsewhere, with longer-staying individuals on Praslin and Silhouette. This species colonised Mahé in the 1990s.

**IDENTIFICATION:** A very large heron, readily identified by its size. The grey upperparts, long neck and black head plumes are distinctive, and in flight its stately wing-beats with bowed wings, long trailing legs and head tucked back on its long neck are characteristic. **ADULT:** The upperparts are grey, with a darker tail and blackish flight feathers. The long neck and underparts are whitish but have blackish streaks up the centre of the throat and breast and there is a black 'shoulder' marking and line that extends down the sides of the chest and along the flanks, which is most noticeable in flight. The black head plumes start from above the eye and project well behind the head; the crown is pale. The bill is very large and dagger-like and is variably greenish-yellow to bright orangey-pink. The long legs are greenish, although they may become bright pink during the breeding season. The eyes are pale. **JUV:** Much as adult, although generally darker and more uniformly grey on the underparts, nape and head, and with a duller bill **(2j)**. **VOICE:** The call is a harsh *"grwnk"*, often given in flight. **BEHAVIOUR:** Hunts for prey along open shorelines, particularly on mudflats, but also along streams and at marshy pools. Birds roost communally in trees around the reclaim area of eastern Mahé. **BREEDING:** Birds are active at their breeding colonies for much of the year. Colonies move as trees fall or nesting sites are lost to development.

## V Purple Heron 3

*Ardea purpurea* Floranten Mov

Length: 80 cm/31 in Wingspan: 125 cm/49 in

A scarce migrant that is probably overlooked because of its habit of skulking in densely vegetated marshes.

**IDENTIFICATION:** A large, slim heron, best separated from the similar Grey Heron by its overall smaller size, slimmer appearance, and much darker, browner plumage. **ADULT:** The dark grey upperparts are washed purple-brown, the tail is dark and the flight feathers are blackish. The long, slim neck and sides of the head are rusty brown and bear black stripes. There is a white, black-streaked line down the throat, neck and chest, but the underparts and underwing coverts are a rich russet. The very long, slim bill is variably yellow, the legs are long and yellowish, and the eyes are pale. **JUV:** More uniformly brown than adult, with mottled ochre-brown upperparts and a less distinct pale throat stripe. They appear dark brown in flight, with darker, blackish flight feathers. **VOICE:** The call is a harsh *"crek"*, normally given in flight. **BEHAVIOUR:** Hunts for prey in vegetated marshes, often standing motionless for long periods with the bill pointing upwards.

1n

1

2j

3

2

3

## Moorhen
1

*Gallinula chloropus* Pouldo

Length: 30 cm/12 in

A fairly common and widespread resident on most islands, inhabiting freshwater wetlands and mangroves. It is wary on the larger islands but often very tame on the offshore islands.

**IDENTIFICATION:** A familiar wetland bird, which wades and swims, and returns furtively to cover when approached. The dark grey and white plumage of the adult, with a red frontal shield and longish orange-yellow legs are diagnostic. **ADULT:** Largely dark charcoal-grey, browner on the back, with a conspicuous broken white bar along the flanks. The tail is short and often held erect, when the white undertail coverts are particularly conspicuous. The red frontal shield is larger in males and the bill is red at the base and has a yellow tip. The legs are variably yellowish green, becoming brighter and more orange during the breeding season, often with a conspicuous red band near the top. The feet are large and lack webs, and the eyes are red. **JUV:** Paler and browner than adult, with a whitish throat and upper breast and a paler face with a darker line through the eye. The head lacks the frontal shield of adults and the bill is dark grey. The legs are dull grey-green **(1j)**. **VOICE:** Gives a range of chucking notes, often a single *"kek"* or *"kek–kek"*, but also a more explosive *"kuk-uk"*. **BEHAVIOUR:** Struts and wades at wetland edges, constantly jerking its head and flicking its tail. Swims readily but flies only short distances during the daytime. Birds can live in extended family groups, defending their territory vigorously against intruders. **BREEDING:** Breeds mostly during the wetter northwest monsoon (October–March), although nests are found year-round. On occasion, several females lay eggs in a collective nest.

## v Common Snipe
2

*Gallinago gallinago* Bekasin Lanmar

Length: 25 cm/9·8 in Wingspan: 42 cm/17 in

A scarce near-annual migrant (October–April), preferring freshwater marshes and flooded grassy areas.

**IDENTIFICATION:** A furtive, mid-sized, brown, shorebird, which is often flushed before it is seen, rising with a characteristic zigzagging flight and distinctive call before dropping back into cover. It has a long, straight bill, warm brown colouration with a striped head, and bold buff mantle stripes. These features separate it from all but the rarer vagrant snipe species. However, the white trailing edge to the dark wing in flight is diagnostic. **ADULT:** Generally rich brown on the upperparts, with bold black and buff stripes on the head and pale buff lines down the mantle. The flanks are dark with grey barring and the belly is whitish. In flight, it shows a distinctive white trailing edge to the largely dark wings. The long, straight bill is dark grey near the tip but paler grey at the base. The legs are mid-length and greenish-grey. **JUV:** As adult. **VOICE:** One or two rasping *"ketch"* are calls given when flushed. **BEHAVIOUR:** Probes in mud with jerky, mechanical movements and will 'freeze' when threatened. When flushed, birds rise with a distinctive zigzagging flight.

## m Garganey
3

*Anas querquedula* Sarsel Ordiner

Length: 39 cm/15 in Wingspan: 63 cm/25 in

A scarce annual migrant (October–April) to wetland pools and marshes.

**IDENTIFICATION:** This is the only duck that occurs regularly in Seychelles. It is a small, dark brown duck and readily takes flight from open water, revealing a dark greenish speculum and broad white trailing edge to the wing. It can also disappear into marsh-side vegetation. Most birds seen in Seychelles are juveniles, although there are records of drakes in breeding plumage in March/April. **ADULT:** Males **(3m)** are unmistakable, with a broad, white supercilium, pinkish-chestnut head and breast, pale grey flanks and, in flight, bright pale blue wing coverts. The bill is dark and quite long and stout, and the legs and feet are blackish. Females **(3f)** are largely mid- to dark brown, with a dark line through the eye and a second dark bar across the cheek from the base of the bill, and a pale loral spot. Greyish webs to the primaries make the wings look pale in flight. **JUV:** Very similar to females. **VOICE:** The typical call is a clear nasal quack, although males have a dry cackling call. **BEHAVIOUR:** Fairly shy and unobtrusive, feeding in marshy pools or at the fringes of more open wetlands, such as La Gogue Dam on Mahé.

1
1j
2
3f
3m

## m Greenshank

1

*Tringa nebularia* Kavalye Lapat Ver

Length: 32 cm/13 in Wingspan: 58 cm/23 in

An annual migrant (August–April) in small numbers, frequenting open shores.

**IDENTIFICATION:** A mid-sized, strikingly pale grey shorebird with a mid-length, slightly upturned bill which has a blue-grey base and a darker grey tip, and diagnostic pale greenish legs. It is smaller, greyer and paler than the superficially similar Bar-tailed Godwit. **ADULT:** In non-breeding plumage, the upperparts are pale and greyish with a faint scaly pattern, whilst the head and neck appear whitish. The underparts are largely white, with some darker streaking across the breast. In flight, the wings appear uniformly dark grey, contrasting with a bold white rump that extends in a 'V' well up the back. The tail appears very pale and is only narrowly barred darker. The bill is relatively thick-based and appears stout and slightly upturned; it is pale blue-grey at the base and dark grey at the tip. The legs are long and normally a pale greyish-green. **JUV:** Darker and browner above than non-breeding adults, with a more heavily streaked breast. **VOICE:** A strident *"tyew tyew tyew"*. **BEHAVIOUR:** Wades elegantly on long legs, often feeding in deeper water than other shorebirds, sweeping its bill from side to side to catch shrimps and small crabs.

## m Bar-tailed Godwit

2

*Limosa lapponica* Limoza Lake Are

Length: 37 cm/15 in Wingspan: 67 cm/26 in

An annual migrant (October–April) in small numbers to intertidal areas. Most birds are of the nominate subspecies *L. l. lapponica*, although the eastern subspecies *L. l. baueri* has been recorded.

**IDENTIFICATION:** A moderately large brownish-grey shorebird with a distinctive long upturned bill, which has a pink base and a darker grey tip. In flight it has dark-tipped but plain wings contrasting with a white rump. The vagrant Black-tailed Godwit *L. limosa* is similar in size and shape but has striking white wing bars in flight. The superficially similar Greenshank is smaller and greyer, has greenish, rather than black, legs, and a shorter, less upturned bill. **ADULT:** In non-breeding plumage, brownish-grey above, with darker feather centres giving a streaky effect, and whitish below with darker smudging across the breast. Fairly prominent whitish supercilia contrast with the darker streaked crown. In flight, the absence of wing-bars on the plain, darker-tipped wings and longish, upturned, bi-coloured bill are distinctive. The legs appear relatively short and are blackish. In the nominate subspecies *L. l. lapponica*, the underwing and rump are white, and the short tail is barred black and white. The eastern race *L. l. baueri* has dark-barred underwings, a dark lower back and upper rump, and a heavily barred brown and white lower rump and tail. **JUV:** Darker above than non-breeding adult, with buffy edges to the wing feathers, and a warm buffy wash across the foreneck and breast. **VOICE:** A nasal *"kewee"*, given in flight. **BEHAVIOUR:** Probes deep into mud of open inter-tidal areas, alone or in small, loose groups.

## m Whimbrel

3

*Numenius phaeopus* Korbizo Ordiner

Length: 40 cm/15 in Wingspan: 83 cm/33 in

A common migrant to open shores and adjacent areas of short grass, with many birds present all-year.

**IDENTIFICATION:** A moderately large brown shorebird, with a distinctive long, down-turned, greyish bill, and strongly marked head, with pronounced dark stripes through the eye and over the sides of the crown, with a pale central crown stripe. The much scarcer Eurasian Curlew is larger and paler, with an obviously longer bill and rather plain head. **ADULT:** Generally brownish with dark streaking and a pale belly, but appears quite dark from a distance. The dark crown has a pale central stripe and there is a prominent dark stripe through the eye, contrasting with pale supercilia. In flight, the prominent white rump contrasts with the relatively plain but dark-tipped wings. The legs are mid-length and bluish-grey, and the bill is long and strongly decurved and uniformly dark grey. **JUV:** As non-breeding adult, although the plumage appears neat and fresh. **VOICE:** A whinnying *"puupuupuupuupuupuu..."*, normally given in flight. **BEHAVIOUR:** Probes deep into substrate, alone or in small, loose groups, and is generally quite wary. It has shorter wings and flies with more rapid wing-beats than Eurasian Curlew.

1n

2n

3n

4n

## Eurasian Curlew
4

*Numenius arquata* Gran Korbizo

Length: 52 cm/21 in Wingspan: 98 cm/39 in

A scarce but annual migrant (October–April) to open shores, where it is heavily outnumbered by the similar Whimbrel. Most birds are of the nominate subspecies *N. a. arquata*, although some show features of the eastern subspecies *N. a. orientalis*.

**IDENTIFICATION:** A large brown shorebird with a very long, down-turned, greyish bill, often showing a pinkish base. It is larger, paler and longer-billed than Whimbrel with a much plainer head. **Adult:** Generally warm grey-brown, with darker streaks and spotting and whiter underparts. The head lacks the bold markings of Whimbrel. In flight, the underwings appear pale. Birds of the eastern subspecies *N. a. orientalis* have unmarked white underwings. The tail and lower rump have dark barring, whilst the upper rump is white. The wing-tips are darker than the rest of the wing but overall ths species appears paler than Whimbrel in flight. The long legs are bluish-grey, and the very long decurved bill normally shows a paler pinkish base and a darker grey tip. **Juv:** As non-breeding adult, although more finely barred on the sides of the breast. **VOICE:** A diagnostic far-carrying fluty whistle "*curlii*". **BEHAVIOUR:** Probes deep into substrate with its long bill for crabs and worms, feeding alone or with other shorebirds but is generally wary.

## m Greater Sandplover

1

*Charadrius leschenaultii*

Gran Plovye Ordiner

Length: 21 cm/8·3 in  Wingspan: 55 cm/22 in

A more numerous and more widespread annual migrant (October–April) than Lesser Sandplover to intertidal areas and adjacent patches of short grass.

**IDENTIFICATION:** Very similar to Lesser Sandplover, but larger, with longer and paler legs and a proportionately longer and heavier bill. The neck appears longer and the head smaller and more angular, often with a flatter crown than Lesser Sandplover. All these features combine to suggest a more elegant appearance than the latter species. These differences are best seen when the two species are viewed together. **Adult:** In breeding plumage, the brick-red breast-band is less extensive than on Lesser Sandplover, but this plumage is rarely seen in Seychelles. Non-breeding birds are mid-sandy brown above and whitish below with a narrow sandy-brownish, normally complete, breast-band. There is an indistinct pale supercilium, and a bold whitish wing-bar in flight. The legs appear quite long and are usually greenish grey, and the bill is relatively short and dark grey. **Juv:** As non-breeding adult, but with broad yellowish fringes to the wing coverts and generally a richer brown colour on the upperparts.

**VOICE:** Generally silent, but may utter a trilling *"piiriiriip"*.

**BEHAVIOUR:** Feeds on open mud or sand in the manner of a Grey Plover *(page 48)*, with slow, elegant strides interspersed with deliberate pecks. The body is usually held horizontally, in contrast to the hunched, angled posture of the Lesser Sandplover.

## m Lesser Sandplover

2

*Charadrius mongolus*  Pti Plovye Ordiner

Length: 18 cm/7·1 in  Wingspan: 50 cm/20 in

An annual migrant (October–April) in small numbers to intertidal areas.

**IDENTIFICATION:** Very similar to Greater Sandplover, but smaller with distinctly shorter and darker legs (the upper leg is often not visible), and a shorter, less heavy bill. They often appear rather short-necked with a large, rounded head. **Adult:** In breeding plumage, the bold brick-red breast-band is thicker and more extensive on the flanks than in Greater Sandplover, but this plumage is rarely seen in Seychelles. Non-breeding birds are mid-sandy-brown above and whitish below, with a sandy-brown, often incomplete, breast-band. The supercilium is indistinct but pale. A bold, whitish wing-bar is visible in flight. The legs appear quite short and are usually blackish grey, and the bill is relatively short and dark grey. **Juv:** As non-breeding adult, except for the presence of paler fringes to the feathers on the upperparts.

**VOICE:** Generally silent, but may call a harsh *"chirrip"*.

**BEHAVIOUR:** Feeds on open mud or sand, making short walks interspersed with deliberate pecks. These movements are rapid and often made with the head hunched up into the body.

## m Ringed Plover

3

*Charadrius hiaticula*  Plovye Kolye Nwanr

Length: 19 cm/7·5 in  Wingspan: 38 cm/15 in

A scarce annual migrant (October–April) to intertidal areas.

**IDENTIFICATION:** A small, robust shorebird, easily identified, if seen well, by its characteristic dark breast-band and facial markings, and bright fleshy-pink to orange legs. **Adult:** Unmistakable in breeding plumage in having a black breast-band and facial markings, contrasting with sandy brown upperparts and white underparts; they also have a distinct white collar around the nape. The legs are shortish and bright orange, and the bill is orange with a blackish tip. In flight there is a conspicuous white wing-bar. Non-breeding adults are duller, having an incomplete breast-band and browner facial markings, an all dark bill and duller pinkish-orange legs. **Juv:** As non-breeding adult.

**VOICE:** The typical call is a soft, fluty *"too-ip"*.

**BEHAVIOUR:** Feeds on open mud or sand, making short walks or runs interspersed with deliberate pecks. Migrants may feed along the sandy paths of smaller islands.

1n
1n
2n
2
3
3n

## m Ruddy Turnstone ①

*Arenaria interpres* Bezros

Length: 22 cm/8·7 in Wingspan: 46 cm/18 in

An abundant migrant to intertidal areas, feeding well inland on smaller islands, with many birds remaining all-year.

**IDENTIFICATION:** A small, stocky, relatively long-tailed shorebird. The boldly patterned face and upperparts in flight, and thick-based bill and bright orange legs are distinctive. **ADULT:** In breeding plumage, males are very boldly marked, with a distinctive black facial pattern on a white background. The mantle and upperwing coverts have chestnut patches on a black background, and the white underparts contrast with the black breast-band. It is even more striking in flight, with the large white rump patch extending well up the back, white wing-bars and a white stripe across the base of each, otherwise blackish, wing. The tail is white with a broad black band near the tip and black uppertail coverts. The female is similarly but less boldly patterned, with a mostly dark crown and face, and reduced chestnut markings. Non-breeding birds are rather uniform dark grey-brown above and white below, with much reduced patterning on the head, and the bold black patches on the sides of the breast having dusky centres. The bill is short and dark and has a thick base tapering to a sharp tip. The legs appear relatively short and are bright orange. **JUV:** Much as non-breeding adult, but pale fringes to the feathers on the upperparts give a scaly appearance. **VOICE:** Gives a variety of calls, including a yelping *"kyew!"*, and a chuckling *"tuk tuk tuk"*. **BEHAVIOUR:** Feeds along open shores, habitually turning over rocks, seaweed etc. to locate small prey; it also forages in the leaf-litter of coastal woodlands on smaller islands.

## m Grey Plover ②

*Pluvialis squatarola* Plovye Sann

Length: 28 cm/11·0 in Wingspan: 60 cm/24 in

A common migrant to intertidal areas and adjacent patches of short grass, with some birds remaining all-year.

**IDENTIFICATION:** A relatively large and squat, predominantly grey shorebird. The black 'armpit' and bold white rump are diagnostic in flight. **ADULT:** The bold breeding plumage consists of black underparts, and black and white spangled upperparts, with white nape, sides of breast and vent. Most birds in Seychelles are in duller non-breeding plumage, which are largely greyish, speckled diffusely blackish above and on the breast, often with a trace of buff. In flight, the blackish flight feathers highlight a distinct whitish wing-bar, the white rump contrasts with the darker upperparts and dark barred tail, and the uniform greyish-white underwings contrast with the diagnostic black 'armpits'. The supercilium is indistinct and pale, with a diffusely darker patch through the eye and paler lores. The bill is quite heavy and black, and the legs are mid-length and dark grey. **JUV:** As non-breeding adult, although often has a distinct buffy wash across the breast. **VOICE:** The call is a diagnostic mellow *"pleeoowii"*. **BEHAVIOUR:** Feeds along open shores and mudflats with slow, elegant strides interspersed with deliberate pecks; it is generally rather alert and shy of approach.

## m Pacific Golden Plover ③

*Pluvialis fulva* Plovye Dore

Length: 23 cm/9·1 in Wingspan: 65 cm/25 in

An annual migrant (October–April) in small numbers to intertidal areas and to areas of short grass, such as airstrips, and mown lawns on Bird.

**IDENTIFICATION:** A mid-sized, rather uniformly grey-brown shorebird. The lack of black 'armpits' and dark (not white) rump separate it in flight from the similar but larger and less elegant Grey Plover. It appears relatively small-headed, long-necked and round-bodied. **ADULT:** The bold breeding plumage consists of black underparts separated from black and gold spangled upperparts by a broad white stripe which is boldest over the eye and down the sides of the neck and breast. Most birds in Seychelles are in duller non-breeding plumage, which are largely grey-brown with richer golden spangled upperparts and paler greyer underparts. In flight it shows an indistinct pale wing-bar, the rump is the same colour as the mantle and tail and the underwings are all grey. The supercilium is short and pale and indistinct, and there are darker patches in front of the eye and on the ear coverts. The bill is mid-length, dark and quite fine, and the legs are relatively long and dark grey. **JUV:** As non-breeding adult. **VOICE:** The call is a diagnostic sharp *"choowit"*. **BEHAVIOUR:** Feeds on open mud and especially on the short turf of airstrips; it is often quite confiding.

m **Crab Plover** 4

*Dromas ardeola* Kavalye

| Length: 40 cm/16 in | Wingspan: 75 cm/30 in |
|---|---|

An annual migrant, occurring in moderate numbers in intertidal areas. It is most numerous from October–April but some birds (mostly immatures) are present all-year.

**IDENTIFICATION:** A very distinctive large black and white shorebird with long legs and an outsized black bill. **ADULT:** The plumage is largely white with a boldly contrasting black mantle and black flight feathers, most prominent in flight. The black bill is massive, and the blue-grey legs are long with disproportionately short toes. The eyes are black. **JUV:** As adult but the wing coverts and tail are grey (not white), the mantle is smudged with dark grey (not cleanly black) and there is a variable amount of black speckling on the hind-crown. **VOICE:** Calls infrequently, a harsh "*quadak*" or a shrill whinnying. **BEHAVIOUR:** Feeds on open mudflats, tidal creeks and along sandy beaches, and is often seen chasing crabs at dusk. It is wary and does not usually allow a close approach, and has a curious running gait, often with the head tucked well into the body. Birds often rest on their haunches.

## m Sanderling

1

*Calidris alba* Bekaso Blan

Length: 20 cm/7·9 in Wingspan: 42 cm/17 in

A fairly common migrant to intertidal areas, particularly sandy beaches. Some birds remain all-year.

**IDENTIFICATION:** A small, very pale grey and white shorebird, which characteristically dashes in and out between waves to probe in freshly wetted sand. The lack of an all-white rump and mid-length, straight bill separate it from Curlew Sandpiper. The Little Stint is even smaller, darker above and has a shorter bill. The absence of a hind-toe is diagnostic. **ADULT:** Birds in breeding plumage, with rusty upperparts are almost never seen in Seychelles. The typical non-breeding plumage comprises very pale grey upperparts and white underparts. The flight feathers are blackish and contrast with a distinct whitish wing-bar. The centre of the tail and rump is blackish with paler sides. The bill is mid-length, relatively straight and black, and the legs are quite short and black. **JUV:** More patterned than non-breeding adults, appearing rather chequered black and white above with dark streaks on the crown and a buffy wash on the sides of the breast. **VOICE:** The call is a liquid *"klit"*, normally given in flight. **BEHAVIOUR:** Typically seen darting back and forth between waves along sandy shores, frequently in small loose groups, and often quite confiding.

## m Little Stint

2

*Calidris minuta* Pti Pti Bekaso

Length: 15 cm/5·9 in Wingspan: 35 cm/14 in

A scarce annual migrant (October–April) to inter-tidal areas. Normally seen singly amongst flocks of other species.

**IDENTIFICATION:** A very small shorebird, darker and browner above than the larger Sanderling, with a short, straight bill and less striking wing-bar in flight. **ADULT:** Non-breeding birds are brownish-grey above, mottled with darker streaks along the centres of the feathers, and blacker on the flight feathers. The underparts are white, though there is a diffuse, often broken, band of streaks across the breast. The crown is dark-streaked, often with a whitish fore-crown and fairly prominent whitish supercilia. In flight, the centre of the rump and tail is dark, as in Sanderling, but the whitish wing-bar is less pronounced. The bill is short, straight and black, and the legs are quite short and black. A hind-toe is present. **JUV:** Much more richly plumaged than non-breeding adult, with a rusty tone above and bold whitish 'tramlines' on the mantle. The wing feathers are dark-centred with pale edgings and the crown is dark-streaked with a prominent pale lateral crown stripe that joins the front of the supercilium. The nape is greyish and there is a rich rusty streaked patch on the breast-sides. **VOICE:** The call is a short *"tit"*. **BEHAVIOUR:** Feeds in a jerky manner and although it may follow waves in the manner of a Sanderling, it is usually less active; it is often quite confiding.

## m Curlew Sandpiper

3

*Calidris ferruginea* Bekaso Korbizo

Length: 20 cm/7·9 in Wingspan: 44 cm/17 in

A common migrant (October–April) to intertidal areas, also feeding on short grass areas, even in the centre of Victoria.

**IDENTIFICATION:** A small, rounded pale grey and white shorebird, with a distinctive long down-curved bill, and square white rump in flight. **ADULT:** Breeding plumage birds with brick-red underparts and boldly speckled upperparts are rarely seen in Seychelles. Non-breeding birds have uniform pale grey upperparts and white underparts, with some light grey streaking on the sides of the breast and a distinct long, pale supercilium. In flight it shows a distinct whitish wing-bar contrasting with blackish flight feathers, and a square white rump, contrasting strongly with the grey mantle and tail. The bill is long, down-curved and black, and the legs are moderately long and black. Birds often appear rather pot-bellied, with a slim neck and head. **JUV:** Similar to non-breeding adult but more scaly-patterned on the upperparts and with a peachy wash across the breast. The crown is streaked darker and there is a more prominent dark line through the eye. **VOICE:** The call is a trilling *"chirrup"*, normally given in flight. **BEHAVIOUR:** Normally feeds in groups, probing its long bill deep into substrate. Flocks often feed on the football fields in central Victoria and on turf around the chalets on Bird Island.

1n

2n
2
2j

3n

## m Wood Sandpiper

1

*Tringa glareola* Sifler Pye Dibwa

Length: 20 cm/5·9 in Wingspan: 37 cm/15 in

A scarce annual migrant (September– April), usually preferring the margins of fresh or brackish marshes to open shores.

**IDENTIFICATION:** A slim, elegant brownish-grey shorebird with a short, straight dark bill, distinctive pale yellowish legs, and a square white rump in flight. It is smaller, darker and shorter-billed than Greenshank *(page 44)*, with a less extensive white rump. **Adult:** Birds in non-breeding plumage are dark brownish-grey above with numerous paler brown spots, and white below with diffuse darker markings across the breast. The bold, pale supercilium extends well beyond the eye and contrasts with the dark crown and dark stripe through the eye. In flight, the absence of wing-bars on plain dark wings, and square white rump is distinctive. The short tail is finely barred black and white. The bill is relatively short, fine and dark, whilst the longish legs are yellowish-green. The toes project well beyond the tail in flight, and the underwing is pale. **Juv:** Darker and browner above than non-breeding adult, with more extensive pale spotting above and a more finely streaked breast.

**VOICE:** A high-pitched quick *"chif if if"* given in flight.

**BEHAVIOUR:** Wades and pecks along the margins of pools. Flies up high quickly when flushed giving its diagnostic call.

## m Terek Sandpiper

2

*Xenus cinereus* Sifler Trakase

Length: 23 cm/9·1 in Wingspan: 58 cm/23 in

An annual migrant (October–April) in small numbers to intertidal areas.

**IDENTIFICATION:** A small grey shorebird with a distinctive steep forehead, long upturned bill, short orangey-yellow legs and a white trailing edge to the wings in flight. The Greenshank *(page 44)* is similar in some respects but is much larger, has longer, greenish legs and a bold white rump in flight. **Adult:** Pale grey above with a black upper edge and carpal area on the closed wing. The underparts are white, with a grey wash across the breast. The forehead is steep, causing the white supercilium to bend up over the dark eye. In flight, the white trailing edge to the otherwise grey wing and uniformly grey mantle, rump, and tail is distinctive. The long, blackish upturned bill has an orangey base, and the legs appear relatively short and are variably dull to bright yellowy-orange. **Juv:** Similar to non-breeding adult, although the black pattern on the closed wing is less pronounced and the legs are duller in colour. **VOICE:** Gives a series of fluty monotone *"weet weet"* notes. **BEHAVIOUR:** Frequently runs with its head held low and bill pointing forward, to pick insects from the surface of the mud. It also habitually bobs the rear of its body like Common Sandpiper.

## m Common Sandpiper

3

*Actitis hypoleucos* Sifler Bat Lake

Length: 19 cm/7·5 in Wingspan: 33 cm/13 in

An annual migrant (September–April) to intertidal areas and to the margins of freshwater bodies such as La Gogue Dam.

**IDENTIFICATION:** A small shorebird, which appears dark above and white below and frequently bobs the rear part of its body when on the ground. In flight it has distinctive flickering wing-beats interspersed with short glides. **Adult:** The upperparts are grey-brown above with some paler scaling. The underparts are clean white, with a contrasting, well-defined dark grey-brown breast marking. The white of the belly reaches above the carpal joint of the closed wing forming a distinctive 'wedge'. The supercilium is straight and pale, contrasting with a long black stripe that runs through the eye; it also has a distinct white eye-ring. In flight, the tail appears quite long, there is a bold white wing-bar across the darker flight feathers, and the rump and tail are the same colour as the mantle. The dark-tipped greenish bill is quite short and straight and the greenish-grey legs appear relatively short. **Juv:** Similar to adult, but more obviously barred on the upperparts, with buffy fringes to the wing feathers. **VOICE:** Gives a characteristic penetrating *"swee wee wee"* when taking flight. **BEHAVIOUR:** Habitually bobs its rear end up and down and walks with a nervous flicking motion. The flickering flight with rapid downbeats followed by short glides is diagnostic.

1n

2n

3n

## m Antarctic Skua

1

*Catharacta antarctica* Poulmov Antartik

Length: 63 cm/25 in Wingspan: 155 cm/61 in

A scarce summer migrant (May–September) to waters around seabird colonies.

**IDENTIFICATION:** A very large, powerful seabird, readily identified by its dark brown plumage and bold white wing flashes. The even rarer South Polar Skua *Catharacta maccormicki* 3 (4 records 1998–2000) is typically less uniformly coloured and usually shows a markedly pale buff nape and straw-coloured underparts, which contrast with the blackish wings. **ADULT:** Typical adults of the subspecies *C. a. lonnbergi* have a uniformly dark-brown body with some paler streaking, especially on the upperparts. The wings are blacker but do not contrast strongly with the body. There is a white patch at the base of the flight feathers, which forms a prominent white flash in flight; this is particularly pronounced on the underwing. At a distance, this species appears all-dark with bold white wing flashes, but at closer range the rather heavy head, neck and bill, and bulky body are evident. **JUV:** Similar to adult, but the plumage is warmer brown overall, often with a reddish-brown cast on the underparts, and with fewer pale streaks.

**VOICE:** Generally silent at sea.

**BEHAVIOUR:** Flies in a rather laboured manner low over the sea on broad blunt wings. It chases and kills smaller seabirds, especially Fairy and Sooty Terns in Seychelles, which are devoured whilst sitting on the surface of the sea. They are frequently mobbed by clouds of terns.

## v Lesser Black-backed (Baltic) Gull

2

*Larus fuscus* Mwet Ledo Nwanr

Length: 52 cm/20 in Wingspan: 125 cm/49 in

A rare migrant to sea coasts (October–March), preferring more sheltered shores.
Most records refer to the nominate subspecies *L. f. fuscus*, but 'Siberian' Gull *L. (fuscus) heuglini* has also recently been observed.

**IDENTIFICATION:** A large, long-winged gull that dwarfs terns and shorebirds. Adults of the nominate form *L. f. fuscus* differ from the form *L. (fuscus) heuglini* (which is treated as a separate species 'Siberian' Gull by some authorities) in their entirely blackish (not grey) upperparts and more limited dark streaking on the white head. Juveniles of the two forms are not safely separable in the field. **ADULT:** The upperparts are entirely blackish, lacking contrasting darker wing-tips and with limited white tips to the primaries. The head, tail and underparts are white, although there is a limited amount of dark streaking on the back of the head. The legs are yellow and the bill is yellow with a red spot on the lower mandible. **JUV:** Generally mottled in appearance, with heavy blackish spotting on the grey-brown mantle and wings and a dark mark around the eye. The underparts are paler, with extensive dark streaking. The wing feathers and tip of the tail are blackish, with a strongly contrasting whitish rump and uppertail coverts. The bill is black and the legs pinkish. **IMM:** Adult plumage is attained over a period of four years and there is therefore a transition through various immature plumages: the blackish mantle and white underparts are apparent from the second year; in the third year the black-tipped tail feathers are replaced by white feathers.

**VOICE:** Generally silent in Seychelles.

**BEHAVIOUR:** The flight is strong and involves much gliding. Birds often perch on mud-banks or on man-made features close to the shore.

1
3
1
2
2j
2
2j

## m Gull-billed Tern ①

*Geochelidon nilotica* Dyanman Gro Labek

Length: 38 cm/15 in Wingspan: 80 cm/31 in

A scarce annual migrant (September–December) to inshore coasts.

**IDENTIFICATION:** A mid-sized chunky, almost gull-like tern, with a relatively short, thick black bill and grey rump. Winter-plumaged birds have a distinctive black mask through the eye on an otherwise white head. The wings are thicker-based and the tail shorter than in Common Tern *(page 58)*. **Adult:** Non-breeding adults have rather uniformly grey wings, mantle rump and tail, and are white below. The head is white with a black mask through and behind the eye, which does not meet across the nape. The eye, bill and rather long legs are black. **Juv:** Similar to non-breeding adult, but with variable amounts of dark scaling on the mantle and wing coverts. The black patch on the ear coverts is less pronounced and there is fine dark streaking on the hind-crown.
**VOICE:** The flight call is a characteristic nasal rising *"ker-vek"*.
**BEHAVIOUR:** Does not normally plunge-dive like other similarly plumaged terns, but hawks insects in the air.

## m Lesser Crested Tern ②

*Sterna bengalensis* Pti Golan Sardin

Length: 36 cm/14 in Wingspan: 80 cm/31 in

A scarce annual migrant (November–April) to inshore coasts, where it is greatly outnumbered by the more widespread Crested Tern.

**IDENTIFICATION:** A mid-sized tern, superficially similar to Crested Tern, but separated from that species by its smaller size, more elegant shape, slightly darker mantle and upperwing, an orange (not yellow) bill, and more deeply-forked tail. **Adult:** The wings, mantle, rump and tail are uniformly smoky grey, and the underparts are white. Breeding birds show an all-black crown but on non-breeding birds the fore-crown is white, the crested hind-crown is black, and there is a variable amount of black speckling on the mid-crown. The long slim bill is bright orange, and the legs are black. **Juv:** Similar to non-breeding adult, but the bill is yellowish-orange and the wings show a pale grey panel on the wing coverts, which contrasts with the rest of the upperwing.
**VOICE:** The call is a grating *"kerik!"*.
**BEHAVIOUR:** The flight is elegant, on long, slender wings. Plunge dives vertically to catch fish, often associates with other terns.

## m Crested Tern ③

*Sterna bergii* Golan Sardin

Length: 45 cm/18 in Wingspan: 105 cm/41 in

A regular migrant to sheltered inshore coasts. Larger numbers occur during the northwest monsoon (October–March) but some birds remain all-year.

**IDENTIFICATION:** A very large, heavy-looking tern with long, slim wings, readily identified by its large pale yellow bill, white fore-crown, ragged black-crested hind-crown and pale grey upperparts. **Adult:** Non-breeding adults have very pale grey wings lacking darker tips, with a uniformly pale grey mantle, rump and tail, and white underparts. The fore-crown is always white and the crested hind-crown is black with variable black speckling on the mid-crown. The large and long bill is dull yellow and the legs are black. **Juv:** Similar to non-breeding adult, but the bill is often a greeny-yellow. Very young birds have variable amounts of black scaling above and blacker primaries.
**VOICE:** The flight call is a harsh *"kri-ek!"*, though immatures sometimes follow adults and give a high-pitched whinnying call.
**BEHAVIOUR:** Has a graceful flight on long, slim wings. Birds frequently perch on sand-bars.

1n

1n

2n

2n
2

3n

3j
3n

## V Black-naped Tern 1

*Sterna sumatrana* Dyanman Likou Nwanr

Length: 31 cm/12 in Wingspan: 60 cm/24 in

Probably a near-annual vagrant (September–March), preferring clear, sheltered coastal waters.

**IDENTIFICATION:** A mid-sized, very pale grey tern, most easily confused with the larger Common Tern or even a distant Fairy Tern *(page 60)*, but separated by its neater and slimmer black band running from behind the eyes to join at the nape, much paler upperparts, pink-tinged underparts, and different feeding behaviour. **ADULT:** Appears mostly white in the field, with a black line that starts just in front of the eye and broadens behind the eye to join across the nape. The mantle and wings are very pale grey with a black edge to the outer primary. The rump and deeply forked tail are white. The bill is medium length, very slim and black, and the legs are black. **JUV:** Similar to adult, but the crown is faintly speckled brown, the wings and mantle have some brown speckling, the black band across the nape is incomplete and the tail lacks streamers. Very young birds have a brownish saddle, which is speckled grey and buff, and a shorter thicker dusky-yellow bill. **VOICE:** The call is a short, sharp repeated *"kik kik"*. **BEHAVIOUR:** Skims low over the sea, snatching prey from surface in the manner of a noddy.

## m Common Tern 2

*Sterna hirundo* Dyanman Lerop

Length: 35 cm/14 in Wingspan: 75 cm/30 in

A regular and widespread migrant (October–April) frequenting more sheltered coastlines.

**IDENTIFICATION:** A mid-sized, pale grey tern, separated from the similar Roseate Tern, by its darker grey upperwing and mantle, shorter forked tail and longer wings. Most birds seen are in non-breeding plumage with a white forehead and black bill. They superficially resemble the smaller and rarer Black-naped Tern but lack the distinctive black head markings of that species. **ADULT:** Birds in breeding plumage have mid-grey wings, white rump, tail and underparts, a complete black cap, and a red bill. In non-breeding plumage, the wings show a prominent blackish carpal bar and blackish-tipped primaries, a distinct white patch develops on the forehead, the long tail streamers are lost and the bill turns black. The legs are relatively short and red. **JUV:** Similar to non-breeding adult but more strikingly patterned with bolder wing markings including a pale grey patch in the centre. The tail is often worn, appearing black-edged and short. **VOICE:** The normal flight calls are a sharp *"chip"*, and the alarm call is a drawn-out *"kree-aaah!"*. **BEHAVIOUR:** The flight action is more languid and graceful than Roseate Tern, and it often hovers and plunge-dives for fish.

## Roseate Tern 3

*Sterna dougallii* Dyanman Roz

Length: 34 cm/13 in Wingspan: 72 cm/28 in

A rare breeder, formerly more widespread in Seychelles but now confined to Aride (*ca.* 1,300 pairs), where it is present April–September. Occasionally breeds on Booby Island and a very scarce migrant elsewhere. The birds in Seychelles are of the subspecies *S. d. arideensis*, which is confined to the western Indian Ocean.

**IDENTIFICATION:** A mid-sized pale grey tern, most easily confused with migrant Common Tern, but generally paler above, with pink-flushed underparts, a more deeply forked tail and longer tail streamers. It often appears rather short-winged and long-tailed in flight. **ADULT:** The upperwing and mantle are very pale grey and the underparts are white, tinged with pink. At a distance, breeding adults appear white, except for a neat black cap and narrow blackish edge to the outer wing. The tail is white and deeply forked with very long tail streamers. The bill is relatively long and slim; it is black early in the breeding season but reddens from the base as the season progresses, becoming completely bright red by July. The legs are relatively long and bright red. **JUV:** Rather different from adult, with a bold black and brown scaly pattern on the mantle, a variably dark bar along the leading edge of the wing and dark centres to the primaries. Unlike juvenile Common Tern, the black cap is almost complete, with a generally dark forecrown, and the legs and shorter bill are blackish. **VOICE:** The fight call is a distinct sharp *"chivik"*, which cuts through the mass of other tern calls on Aride. It also produces a harsh *"kraak!"* call. **BEHAVIOUR:** The flight is direct and level on fast and shallow wing-beats. It hovers less than other similar terns, circling and stalling at moderate height before plunge-diving. Birds are often seen diving frenetically through surface-feeding aggregations of Lesser Noddy. They roost on coastal rocks. **BREEDING:** Breeds in dense colonies at traditional sites, where single-egg clutches are laid in May–June; the fledglings depart by the end of August. The main colony on Aride is, uniquely, in dense woodland.

1
1
2n
2n
3
3
3j
3c

## Fairy Tern

(1)

*Gygis alba* Golan Blan

Length: 30cm/12in Wingspan: 80cm/31in

A widespread resident, breeding in large numbers on smaller rat-free islands, and in smaller numbers on all the larger islands, even in shade trees in central Victoria.

**IDENTIFICATION:** A small to mid-sized tern, readily distinguished by its completely white plumage and slightly upturned blue-based black bill. At a distance, it is separated from other whitish terns by the lack of black markings on the head, its weak and indirect flight on short wings and noddy-like feeding actions. **ADULT:** Uniformly white, with relatively short and pointed wings and a shortish, bluntly forked tail. The mostly black bill has a cobalt-blue base, is slightly upturned and is rather tapered. The legs and feet are grey-blue. The eyes are black and appear unusually large. **JUV:** Differs from adult in having a black spot behind the eye, brownish mottling on the nape, greyer upperwings, black shafts to the outer primaries, brown tips to the feathers on the mantle and an all-black bill. **VOICE:** Perching birds emit a soft repeated series of *"yerp"* notes. The alarm call is a bouncing *"gung gung gung"*. **BEHAVIOUR:** The flight seems weak and fluttery, with rapid wing-beats. It dips to snatch small fish from the surface of the sea in the manner of Lesser Noddy. **BREEDING:** Usually nests in loose colonies. Breeding occurs year-round, although there is a peak of egg-laying in November–January. The single egg is laid directly onto the branch of a tree.

## (m) White-winged Tern

(2)

*Chlidonias leucopterus* Dyanman Lezel Blan

Length: 22cm/8·7in Wingspan: 53cm/21in

An annual migrant (September–April) in very small numbers to sheltered coasts and freshwater bodies, sometimes seen hawking insects over nearby areas of short grass.

**IDENTIFICATION:** A small, grey-backed tern, most easily identified by its relatively short, broad-based wings, limited black head markings and characteristic swooping and dipping flight action. **ADULT:** Breeding birds are boldly marked black and white but most birds seen in Seychelles are in non-breeding plumage. The mantle and upperwings are mid-grey, usually with diffusely darker wing-tips and darker bars along both the leading and trailing edges of the wings. The underparts and rump are white, the latter contrasting with the grey mantle and tail. The forehead is white, but there is a black tear-shaped mark running through and behind the eye and there is a small diffuse black patch on the hind-crown, contrasting with the whiter nape. The tail is short and only slightly forked. The bill is short, relatively stocky and black, and the legs are dull orange. **JUV:** Similar to non-breeding adult, but the dark cap is more extensive and the mantle is a darker brownish-grey (this gradually moults to grey) forming a distinct 'saddle', the rump is greyer, and the upperwing pattern is a darker grey-brown. **VOICE:** Generally silent. **BEHAVIOUR:** The dainty, buoyant flight involves much swooping and dipping to the surface of the water or mud to pick off insects, though birds also hawk insects in mid-air.

## (m) Saunders' Tern

(3)

*Sternula saundersi* Pti Dyanman Saunders

Length: 24cm/9·4in Wingspan: 52cm/20in

An annual migrant (October–April) in small numbers to coasts, preferring more sheltered waters.

**IDENTIFICATION:** A very small grey-backed tern, readily separated from all other terns by its tiny size and mid-grey upperparts with a pronounced blackish leading edge to the wing-tips. The very similar Little Tern *S. albifrons* is an extremely rare vagrant. **ADULT:** Most birds seen are in non-breeding plumage. The mantle and upperwings are mid-grey with a bold black leading edge to the outer half of the wing and black wing-tips forming a distinctive black wedge. The rump and tail are ash-grey and the underparts are white. The forehead and lores are whitish, contrasting with a blackish line that runs from the eye and merges with a blackish hind-crown. The relatively short, slim pointed bill is black, and the legs are yellowish-brown. **JUV:** As non-breeding adult, but the fore-crown is tipped brown and the hind-crown patch is browner and less extensive. The grey of the mantle and upperwing is variably tipped brown. **VOICE:** The call is a short sharp *"kik kik"*. **BEHAVIOUR:** Hovers with very rapid wing-beats on relatively long, narrow, pointed wings, and plunge-dives directly into the sea for small fish. Roosts on sand-bars and mudflats.

1c
1

2
2
2n

3n
3n

## Bridled Tern ❶

*Onychoprion anaethetus* Fansen

Length: 36 cm/14 in Wingspan: 76 cm/30 in

A widespread resident, breeding on smaller rat-free rocky islands, and ranging widely to feed along coasts. It is present year-round.

**IDENTIFICATION:** A mid-sized tern with dark upperparts and white underparts. It is most easily confused with Sooty Tern, but is considerably smaller and slighter with a more elegant flight; it is also variably dark grey-brown above, rather than black, has a longer white stripe over the eye, and a proportionally smaller, rounded head. **Adult:** In flight, the mantle appears paler and greyer than the dark brown upperwing, rump and deeply forked tail. The crown and bold line through the eye are black, contrasting with a narrow white forehead and thin white line over the eye. The underparts are white. The nape is often paler than the rest of the upperparts. The tail shows thin white outer tail feathers in flight, but lacks the streamers of adult Sooty Terns. The bill is relatively short and dark, and the legs are black. **Juv:** Similar to adult, but there is a grey-brown wash to the sides of the breast, the head pattern is less clearly defined and the legs are pinkish-grey. Some birds show pale scaling on the upperparts.

**VOICE:** The typical call given by birds at the breeding colonies is a yapping *"kwup kwup"*, although they also give a plaintive *"kaar"* and a high-pitched trilling *"prrrrr"*. Juveniles emit a high-pitched, plaintive, rolling *"prrring!"*.

**BEHAVIOUR:** The flight is lighter and more graceful than Sooty Tern, birds swooping low over the surface of the sea to catch small fish in a similar manner to noddies, with which it often feeds. It roosts both on coastal rocks and in tall beach-crest trees such as *Casuarina* trees.

**BREEDING:** Breeds at low density in loose colonies, with males guarding nest-sites located under boulders or in crevices in cliffs. It has an unusual 8–9 month breeding cycle, with single-egg clutches being laid largely synchronously across the islands.

## Sooty Tern ❷

*Onychoprion fuscata* Golet

Length: 44 cm/17 in Wingspan: 90 cm/35 in

An abundant seabird, breeding in hundreds of thousands on Aride and Bird Island, with smaller colonies on Recif, L'Ilot Frégate and Booby Island. It is widespread in coastal waters during the breeding season (April–October) but very few birds remain all-year.
Colonies are subject to intensive, illegal poaching of both eggs and adult birds, and many have been lost in recent decades.

**IDENTIFICATION:** A large, strikingly piebald tern with black upperparts and white underparts. The boldly forked tail has long white-edged tail streamers. The Bridled Tern is superficially similar but is rather smaller, with a relatively smaller head and lighter flight, and is grey-brown, not black, above. **Adult:** Strikingly black above with white underparts that often appears pale blue in strong light. The wings, mantle, rump and forked tail are uniformly black, whilst the long tail streamers are broadly edged white. The crown is black and there is never a pale collar as in Bridled Tern. The white forehead is broad but does not extend far over the eye. The long, black bill is quite broad-based, and the legs are black. **Juv:** Very different from adult, being largely blackish-grey, except for the lower belly, with rows of white spots forming bars across the upperparts and wings. In flight, the whitish underwing and white lower belly contrast with the otherwise dark plumage.

**VOICE:** The typical call is a diagnostic *"kee-arr-you!"* or variants thereof. The alarm call is a grating *"krroww!"*. Young juveniles emit a series of high-pitched *"wik wik wik..."* calls, whilst fledglings give a scratchy, higher-pitched version of the adult call *"weet-ee-weet"*.

**BEHAVIOUR:** Flies strongly on long, narrow, pointed wings, often gliding high, but also swooping low over the sea to catch small fish, often in dense feeding flocks. It habitually mobs predatory skuas, and also injured birds.

**BREEDING:** Breeds in dense colonies, usually on open rock or sand, but uniquely on Aride it nests beneath the canopy of dense hill woodland. The clutch of one or two eggs is laid in May–June, with chicks fledging in late August–October.

1

2

1j

2j

1

2

## Lesser Noddy

1

*Anous tenuirostris* Kordonnyen, Kelek

Length: 32 cm/13 in Wingspan: 60 cm/24 in

A numerous resident seabird, breeding in large numbers on smaller rat-free islands, and commonly seen around all islands.

**IDENTIFICATION:** A mid-sized, all-dark tern with a whitish fore-crown merging into a pale grey crown. It is most easily separated from the similar Brown Noddy by its smaller size, slighter build, shorter tail and wings, and more rapid flight action. At close range, the plumage is markedly darker than Brown Noddy and lacks any contrast in the wing, the pale grey of the crown extends further onto the nape, and the bill is much finer. Both species have a well-marked pale, broken eye-ring above and below the eye. **ADULT:** The plumage is uniformly dark sooty-grey above and below, relieved only by a greyish white fore-crown and crown, which is more extensive than in Brown Noddy, extending below the front of the dark eye, giving the face a soft expression. The black bill is long and needle-like, and the legs and feet are blackish. The legs are shorter and weaker than in Brown Noddy. **JUV:** Similar to adult, but the bill is relatively shorter and thicker, the whitish crown patch more diffuse, and there are paler brown edges to the wing coverts forming faint bars.
**VOICE:** The normal call at the breeding colonies is a frequently uttered rolling *"kirriik"*, but birds also give a more complaining *"krrrrak!"*. Juveniles in the nest emit plaintive piping notes.
**BEHAVIOUR:** The flight appears weak, with rapid beats of the short, pointed wings. It dips to snatch small fish from the surface, and often feeds in dense flocks when conditions are favourable. Birds regularly rest on the sea. On land, it walks less easily than Brown Noddy, but habitually bows its head (or 'nods') when approached by another noddy.
**BREEDING:** Often breeds in large, noisy colonies, each pair producing a single egg, which is laid in a flimsy nest of leaves placed on the branches of a tree. The onset of breeding varies depending on local feeding conditions, but is typically April–June. Small numbers nest during the northwest monsoon (December–February).

## Brown Noddy

2

*Anous stolidus* Makwa

Length: 43 cm/17 in Wingspan: 83 cm/33 in

A widespread resident seabird, breeding in moderate numbers on smaller rat-free islands, and commonly seen around all islands.

**IDENTIFICATION:** A large dark tern with a whitish fore-crown and pale grey crown. It is larger and more heavily built than the similar Lesser Noddy. The tail and wings are conspicuously longer, and the flight action more graceful and relaxed. At close range, the plumage is paler and browner than Lesser Noddy, with contrastingly dark flight feathers and tail, and a sturdier, thicker-based bill. Additionally, the grey of the crown does not extend onto the nape, and the lores are dark, giving a more fierce facial expression. It is separated from distant juvenile Sooty Terns *(page 62)* by its pale crown, and dark (not white) vent. **ADULT:** The body is uniformly dark brown (paler in strong light), but the underwing is distinctly paler, and the upperwing shows a contrast between the darker primaries and secondaries and the paler wing coverts. The tail feathers are also darker than the uppertail coverts and rump. The fore-crown is greyish-white, as in the Lesser Noddy, but is less extensive and does not extend in front of or below the eye. The black bill is thick-based and quite sturdy, and the legs and feet are blackish. A salmon-pink tongue is visible on gaping birds. **JUV:** Similar to adult but the bill is shorter and the whitish crown patch is smaller. Nestlings are very variable in colour, ranging from black, through grey-brown, to fawn and white.
**VOICE:** The normal call at the breeding colonies is a croaky *"makwaa"*, but birds also give a more aggressive *"kyaar!"* and a more plaintive *"keeo"*.
**BEHAVIOUR:** The flight is strong, often gliding on long, angled wings. It dips to snatch small fish from the surface, but also chases smaller terns to steal their prey. Birds generally feed alone or in small groups. The legs are longer and sturdier than those of the Lesser Noddy and it is able to walk easily on the ground. Birds habitually bow their head (or 'nod') when approached by another noddy.
**BREEDING:** A single, blotched egg is typically laid in a nest placed at the base of a coconut palm frond, or on a rock platform, though it sometimes nests high in larger trees, or on bare ground. Different groups of birds breed at different times of year; most begin nesting at the start of the southeast monsoon (April–June), though smaller numbers nest during the northwest monsoon (November–February).

1

2c

1c

2

## E Seychelles Kestrel 1

*Falco araea* Katiti Sesel

Length: 19 cm/7·5 in Wingspan: 42 cm/17 in

A fairly common and widespread resident on Mahé, with smaller numbers on Silhouette and North Island. It is very scarce on Praslin (Grand Anse to Consolation and above Anse Boudin), extirpated from La Digue and absent elsewhere.

CONSERVATION STATUS: **Vulnerable**

**IDENTIFICATION:** A very small falcon, which sits for long periods on exposed perches such as telegraph wires, trees, or the roofs of buildings, before darting to catch prey, such as day geckos, small birds or insects. It is the only falcon likely to be seen, and is readily identified by its tiny size, rich brown upperparts, buffy underparts and narrow, banded tail. **ADULT:** The male has a neat mid-grey crown and short, dark moustachial stripes. The female has a dark brown crown. Both sexes have rich chestnut upperparts, which are spotted black, and unstreaked buff underparts with a clean white throat. In flight, the wings appear quite narrow and short. The short, narrow tail is grey with a thin pale tip and has black bars, including a thick, black sub-terminal band. The legs, cere and base of the bill are yellow, and the eyes are dark. **JUV:** Similar to adult female, but more russet generally, with dark barring across the mantle and wings, and some dark barring and spotting on the flanks and belly. It has a prominent black 'tear-line' through the eye. The legs are yellow, as in adults, but the cere and base of the bill are bluish. **VOICE:** The call is a thin but piercing *"titititi"*, often given by juveniles. **BEHAVIOUR:** Sits for long periods, often on exposed perches, before darting for prey. The flight is rapid on short wings, but it also soars in sweeping circles over its territory. **BREEDING:** Breeds mostly during the northwest monsoon (October–March), laying a clutch of two white eggs. Nests are often located in church towers, as well as in natural cavities in cliffs and trees.

## V European Cuckoo 2

*Cuculus canorus* Koukou Sann

Length: 34 cm/13 in Wingspan: 57 cm/22 in

A rare but probably annual migrant (October–April), preferring open areas with scattered trees. Individuals may stay for weeks within a very small area.

**IDENTIFICATION:** A typical cuckoo, with a long slim body, long rounded tail and longish pointed wings that droop when the bird is perched. This is the most regularly seen cuckoo in Seychelles, although other even scarcer vagrant species have been recorded. **ADULT:** Typical adults have a pale grey head, breast and upperparts, with darker grey flight feathers and tail. The belly and vent are white with bold black barring. The tail is blackish-grey with a buff tip, and the underside is tipped white. In flight, the underwing can be seen to be barred black and white. The legs and feet are orange with darker claws, and the bill is orange with a dark grey tip. The iris is greenish-yellow with a prominent black pupil and yellow eye-ring. Red-phase females differ in having a reddish-brown ground colour to the upperparts, which are finely barred black like the underparts. **JUV:** Similar to adult but the upperparts and breast are darker grey and the feathers are broadly tipped whitish and buff giving a very scaly appearance. There is a small white patch present on the nape and the eye is darker. Red-phase juveniles differ from females in having narrow white feather edges on the upperparts and the white patch on the nape. **VOICE:** Migrant birds are generally silent. **BEHAVIOUR:** The flight action is fast and direct. Birds often perch for long periods.

## V European Roller 3

*Coracias garrulus* Rolye Lerop

Length: 30 cm/12 in Wingspan: 55 cm/22 in

A rare migrant (October–January) to open areas with scattered trees. It has occurred almost annually in recent years, with individuals often staying for several days or weeks.

**IDENTIFICATION:** An unmistakable heavily -built, large-headed pale blue bird with a brown mantle, which habitually perches conspicuously on posts or wires. Most birds seen in Seychelles are juveniles. **ADULT:** The head, upperwing coverts and underparts are bright pale blue, the rump and underwing ultramarine blue, the mantle reddish-brown, and the primaries blackish. There is a blackish line through the dark eye. The tail is quite long and square-ended with pale blue tip. The bill is stout and black, and the feet are pinkish. **JUV:** Similar to adult but generally paler and greyer powder-blue with a dull brown mantle. **VOICE:** Migrants are generally silent. **BEHAVIOUR:** Perches conspicuously on post or wires, flapping to the ground or to the next post on broad wings with a lazy, floppy flight.

1

2

3

## e Seychelles Black Parrot 1

*Coracopsis [nigra] barklyi* Kato Nwar

Length: 30 cm/12 in Wingspan: 50 cm/20 in

A scarce resident (*ca.* 200–300 birds), restricted to wooded habitats on Praslin and in very small numbers on adjacent Curieuse. It is generally treated as an endemic subspecies of the Lesser Vasa Parrot of Madagascar. However, it is a distinctive form which some authorities consider warrants specific status. If this were followed, its conservation status would need to be re-assessed and it would probably be categorised as Critically Endangered.

**IDENTIFICATION:** A uniformly dull grey-brown (not black) mid-sized parrot, unlikely to be confused with any other bird in Seychelles if seen well. Distant flying birds could be mistaken for Madagascar Turtle Dove *(page 70)* or Seychelles Blue Pigeon, but appear sleeker bodied, bullet-headed and longer tailed, lack contrast in plumage tones, and have a more rapid flight. **ADULT:** The upperparts, body, wings, head and tail are uniformly grey-brown. The body is relatively slim and the tail is long. The eye and mid-sized, hooked bill are dark grey, and the feet and legs are black. **JUV:** Similar to adult, but the upperparts appear more mottled with paler edges to the wing coverts, the overall appearance is often more scruffy, and the bill is blacker.
**VOICE:** The typical call is a surprisingly high-pitched and melodious series of whistled notes.
**BEHAVIOUR:** Flies fast and direct with rapid beats of the relatively short, round-tipped wings. It clambers acrobatically along branches and up the trunks of favoured fruiting trees, using both its beak and claws. Birds can be very unobtrusive when they are perched.
**BREEDING:** Nests in natural cavities in trees, as well as in boxes that are provided specifically for their use. Fledged young are seen during much of the year.

## E Seychelles Blue Pigeon 2

*Alectroenas pulcherrima* Pizon Olande Sesel

Length: 24 cm/9·4 in Wingspan: 40 cm/16 in

A widespread and relatively common endemic pigeon, frequenting wooded habitats on all but Bird and the smallest of the granitic islands.

**IDENTIFICATION:** Unmistakable if seen well. The largely dark blue plumage, with contrasting pale grey breast and nape, bright red fore-crown and facial wattles, is diagnostic. It appears largely dark at a distance but in flight has pale patches in the rounded wings. In this respect it may recall Common Myna *(page 84)* but that species is smaller and less dumpy, has bold white wing-patches and a dark head. **ADULT:** The wings, back, tail and belly are uniformly dark blue, contrasting sharply with the pale grey breast, throat, nape and rear crown. The vent is grey with diffuse darker barring. The forehead is bald and red, as are the wattles and areas of bare skin around the orange eye. Pale patches at the bases of the broad, rounded wings are noticeable in flight. The tail is short and square-ended. The bill is short and yellowish, and the feet and legs are dark grey. **JUV:** Recently fledged birds are very different from adult, being largely pale grey with a darker grey head and wings, an olive-grey mantle, and bright golden fringes to the mantle, rump, tail and upperwing coverts. The primaries and primary coverts have broad white fringes. The facial skin is dark grey, whilst the feet and bill are bluish-grey and the iris is brown. They gradually moult into adult plumage, but in immature plumage the crown is pale pink, there are no facial wattles and the vent is whitish.
**VOICE:** The song is an accelerating series of gruff notes which trail off in volume: "*wraugh wraugh wraugh wraugh-wraugh-wraugh-wraugh...*" Otherwise, birds are normally silent.
**BEHAVIOUR:** Males chase each other in rapid flight and have a distinctive display flight, ascending with a clap of the wings and then descending with their wings held stiffly down at a steep angle to the body. Birds may congregate at favoured fruiting trees, but otherwise are generally solitary and surprisingly unobtrusive. They clamber along branches in the manner of a parrot.
**BREEDING:** A single white egg is laid in a simple twig nest built on the fork of a branch. It breeds throughout the year, with fledglings appearing in most months.

1
1

2c
2

## e Seychelles Turtle Dove

1

*Streptopelia picturata rostrata* Tourtrel-de-Zil

Length: 26 cm/10·2 in Wingspan: 37 cm/15 in

This endemic form has largely been replaced by the Madagascar Turtle Dove, but some birds retaining features of the endemic subspecies *S. p. rostrata* still occur on Aride, Bird, Cousin and Cousine.

**IDENTIFICATION:** Separated from the Madagascar Turtle Dove by its smaller size, darker more uniformly brown body and diagnostic reddish-brown head. **ADULT:** Most birds on the islands listed opposite probably represent intergrades between the two forms. The most extreme birds are obviously smaller and darker, with reddish-brown heads, chestnut rather than vinous upperparts, and darker, browner underparts often with a dark vent. The bill is darker and grey, lacking a pale greenish tip. **JUV:** Uniformly dark greyish-brown. **VOICE:** As Madagascar Turtle Dove. **BEHAVIOUR:** As Madagascar Turtle Dove, but often aggressive towards its larger congener and possibly more prevalent in woods away from habitation, at least on Aride. **BREEDING:** As Madagascar Turtle Dove.

## I? Madagascar Turtle Dove

2

*Streptopelia p. picturata* Tourtrel-de-Zil

Length: 28 cm/11·0 in Wingspan: 44 cm/17 in

A widespread resident, occurring in wooded habitats on all but the smallest offshore islands. It is presumed to have been introduced from Madagascar and has largely swamped the endemic Seychelles Turtle Dove.

**IDENTIFICATION:** A squat, dark mid-sized ground-dwelling dove, frequenting wooded areas. The Feral Pigeon *(page 84)* is more of an urban dweller and is slimmer, longer-tailed and occurs in a wide range of plumages, none of which closely resemble this species. **ADULT:** Rather variable, particularly on some of the smaller islands. It is mostly dark vinous chestnut above, with a paler, grey rump and brown wings. The breast is vinous-grey, the belly paler and pinker, and the vent whitish. The head is paler than the mantle and is typically mid-grey, whilst the tail is darker grey, with a distinct paler band on the underside and paler outer corners on the upperside. There is a prominent red eye-ring around the variably orange eye, whilst the pale greenish bill has a fleshy reddish base. The legs and feet are purplish-pink along the front and greyer along the back. **JUV:** Less clearly marked than adult, being mostly greyish-brown, with rusty edgings below, and often appearing rather scruffy. **VOICE:** A deep rolling "*wooh, wooh*". **BEHAVIOUR:** Feeds on the ground in dense woodland, but also on lawns under looser canopies; it is not particularly shy. **BREEDING:** On Aride, most nests are built during the southeast monsoon and contain up to three eggs, the chicks fledging from June–September. It apparently breeds more frequently during the northwest monsoon (November–March) on the larger islands.

## I Barred Ground Dove

3

*Geopelia striata* Tourtrel Koko

Length: 21 cm/8·3 in Wingspan: 25 cm/9·8 in

An abundant and widespread species introduced from south Asia, frequenting more open man-made habitats.

**IDENTIFICATION:** Easily separated from other doves by its much smaller size, black-barred buff upperparts and long, slim white-edged tail. **ADULT:** The upperparts are buff, boldly barred black across the mantle, wing and uppertail coverts. The barring is finer on the paler, greyer sides of the breast and on the flanks. The unbarred centre of the breast is pinkish and the lower belly and vent pale are buff. The long, thin, graduated tail has a bold white edge, which contrasts with a darker centre. The head lacks bars, and the forehead, face and throat are greyer. The broad cere and base of the small, fine blue-grey bill are pale blue, and the feet and legs are pink. **JUV:** Similar to adult, but paler with a shorter tail. **VOICE:** The song is a monotonous but diagnostic cooing, which rises and then falls in pitch: "*ker-ker ker-kerrr coo coo coo coo*". Birds also produce a frog-like croaking "*kkrroww*" and the male a rolling "*waaw!*" during display. **BEHAVIOUR:** Often extremely confiding when feeding on the ground for scraps in open-air restaurants, parks, hotel grounds etc., but shyer away from human habitation. The flight is rapid and bustling on short rounded wings, but birds rarely fly far. Males sing when competing for females, puffing up their breast and fanning their tail. **BREEDING:** Breeds over much of the year, but with a pronounced peak during the wetter northwest monsoon (November–March).

1

2

3

## E Seychelles Scops-owl

*Otus insularis* — Syer — 1

Length: 21 cm/8·3 in | Wingspan: 40 cm/16 in

A rare endemic resident (<350 birds), confined to mid-elevation forests on Mahé.

CONSERVATION STATUS: **Endangered**

**IDENTIFICATION:** A small, dumpy chestnut-buff owl with a squat, rounded head, unlikely to be confused with any other species within its very small range. Beware that the similar Eurasian Scops-owl *Otus scops* has been recorded in Seychelles as a vagrant, but can be distinguished by its ear-tufts. **ADULT:** Rather uniformly russet-buff above and below, mottled with darker streaks and with paler buff spots on the mantle and wing coverts. The flight feathers are barred alternately black and buff. In flight, the wings appear broad, rounded and quite short, whilst the tail is short. The eyes are bright yellow in some individuals and orange-red in others, the small pupils giving the birds a 'wild-eyed' look. The short, hooked bill is black, and the legs are long, unfeathered and yellow. **JUV:** Similar to adult.

**VOICE:** The call is a diagnostic rhythmic rasping *"wargh"* interspersed with knocking notes. Pairs duet with *"wargh"* and *"wurgh"* notes. Juveniles give a higher-pitched *"hichup!"*.

**BEHAVIOUR:** Exclusively nocturnal, calling most frequently during the first two hours after dusk. It is most easily detected by its soft but carrying call. Birds perch on low to mid-level branches in tall forest, but are not shy and can be inquisitive to intruders.

**BREEDING:** The first nest was only discovered as recently as 1999, and it is now known that a single egg is laid in a natural tree hollow. Pairs begin calling in April–May and breeding occurs during the southeast monsoon (May–September).

## I Barn Owl

*Tyto alba* — Ibou Blan — 2

Length: 36 cm/14 in | Wingspan: 90 cm/35 in

A widespread resident, introduced in the 1950s from east Africa in an attempt to control rats. Unfortunately this has resulted in it decimating Fairy Tern populations on all but the smaller offshore islands.

**IDENTIFICATION:** The only large pale owl in Seychelles, appearing largely white if seen in car headlights. The slim body, long round-tipped wings and pale heart-shaped face are unmistakable. Seychelles Scops-owl is much smaller and is extremely cryptic. **ADULT:** The upperparts are tawny-buff with grey spots and mottling. In flight, the wings appear long and rounded, with broad-spaced dark bars across the flight feathers, and the tail is relatively short and barred. The underparts are white but finely and sometimes densely spotted black. The white, heart-shaped face with contrasting black eyes is diagnostic. The short, hooked bill is pink, the legs are long and the lower part of the tarsus and feet are pink. Males often appear smaller and paler than females. **JUV:** Similar to adult.

**VOICE:** The song is a repeated shrieking *"scrrreee!"*, but birds also gives hoarse alarm calls and a softer rolling *"prrrrr"* contact note.

**BEHAVIOUR:** Almost exclusively nocturnal, being most active in the first two hours after dusk when it quarters open areas searching for rodents or slips through trees to prey upon roosting birds. It often uses conspicuous perches, but roosts during day in rocky clefts, high in a tree or even in the eaves of buildings.

**BREEDING:** Nests year-round, laying a clutch of up to eight white eggs on a ledge in a rocky cleft, tree hollow or old building.

1

2

## E Seychelles Swiftlet (1)

*Aerodramus elaphrus* — Zirondel Sesel

Length: 11 cm/4·3 in   Wingspan: 28 cm/11·0 in

An endemic resident restricted to, and easily seen on, Mahé, Praslin and La Digue, although overall numbers may not be large. It is a rare vagrant elsewhere.

CONSERVATION STATUS: **Vulnerable**

**IDENTIFICATION:** A small, dark, slender swift with erratic rapid fluttering flight on short, slim wings. It is unlikely to be confused with any other bird, except vagrant swift and swallow species, particularly on offshore islands, but its small size, short tail, uniform plumage and flight action are distinctive. **ADULT:** Dark sooty grey-brown above, with slightly paler greyer underparts. It often appears uniformly dark in flight, but in good light a thin, pale band across the nape and a more diffuse pale band across the rump extending to the sides of the vent can be seen. **JUV:** Similar to adult.

**VOICE:** Small flocks emit a soft twittering "*chitchitchit*".

**BEHAVIOUR:** An aerial feeder, often feeding high over forested hillsides during the day, but descending to feed low over plateau areas towards dusk or during wet weather. The flight is rapid and erratic on short, slender wings.

**BREEDING:** Nests in a handful of deep caves in the mountains of the larger islands, mostly during the southeast monsoon (May–September). A single egg is laid in a bracket-shaped nest of lichens and saliva glued to the ceiling of the cave.

## V Barn Swallow (2)

*Hirundo rustica* — Zirondel Lerop

Length: 19 cm/7·5 in (including 5 cm/2·0 in tail streamers)   Wingspan: 33 cm/13 in

An annual migrant (October–May), usually seen hawking over open areas such as airstrips or around buildings on offshore islands.

**IDENTIFICATION:** A familiar aerial feeder, separated from Seychelles Cave Swiftlet, by its blue-black upperparts, white underparts, long pointed wings and long forked tail. **ADULT:** The shiny upperparts are rich blue-black and the underparts are white with a blue-black breast band and red forehead and throat. The tail is dark blue and forked, with long tail streamers; it has a white band on the underside, which appears as white spots from above when the tail is fanned. **JUV:** Similar to adult but lacks the blue sheen to the upperparts and tail streamers, and the face and throat are brownish-pink.

**VOICE:** Migrants are generally silent.

**BEHAVIOUR:** Hawks for insects low over the ground or water, making frequent aerial passes with fast clipped wing beats. Birds often perch conspicuously on wires, branches or roofs.

## V Tree Pipit (3)

*Anthus trivialis* — Pipit Pye Dibwa

Length: 15 cm/5·9 in   Wingspan: 26 cm/10·2 in

An annual migrant in very small numbers (October–December), favouring areas of close-cropped grass, such as hotel lawns, playing fields and airstrips.

**IDENTIFICATION:** A small, buffy-brown, dark-streaked bird, which walks across open grassy areas with an elegant gait on longish pinks legs. The slim bill, boldly streaked breast, horizontal stance and longer tail separate it from the similarly-plumaged Madagascar Fody *(page 80)*. **ADULT:** The plumage is warm buffy-brown above and whiter below, with a buff wash across the breast. It is has extensive dark streaking, especially on the mantle, breast and flanks, with a prominent black malar stripe setting off the clear, pale throat. There is a buff line over the eye from the lores. The rump is buffy-brown and unstreaked, and the longish tail is darker brown with whitish outertail feathers, which are prominent in flight. The legs are long and pink, and the dark-edged, pinkish bill is slim and pointed. **JUV:** Similar to adult.

**VOICE:** Migrants are generally silent.

**BEHAVIOUR:** Walks with an elegant gait on the ground, picking for insects. It often flies into a tree if flushed, characteristically pumping its tail downward when perched.

1

2
2m

3

## E Seychelles Paradise Flycatcher

1

*Terpsiphone corvina* — Vev

Length: 20cm/7·9in Wingspan: 23cm/9·1in (+16cm/6·3in tail streamers on male)

A highly attractive endangered endemic flycatcher restricted to the plateau and adjacent higher areas of La Digue (<200 birds), with a tiny relict population on Marianne, and occasional records from Félicité. It was lost from the marshes of eastern Praslin as recently as 1989. The species is threatened by ongoing habitat loss on La Digue, although it is the subject of a recovery programme.

CONSERVATION STATUS: **Critically Endangered**

**IDENTIFICATION:** An unmistakable species. ADULT: Males **(1m)** have an entirely black plumage, which shows a glossy bluish sheen at close range, and is blackest on the head, chin and throat. There is a shaggy crest on the hind-crown, and the central tail feathers are very long, forming trailing streamers. The bill, gape and eye-ring are bright pale blue and the short legs are blue-grey. The eye is black. Females **(1f)** are very different, having a black head and throat contrasting with whitish-buff underparts and rufous mantle, wings and tail. The tail lacks the streamers of the male, but the bare parts are the same colour. **JUV:** Similar to adult female, the males slowly moulting into their black adult plumage and often showing black feathers on the mantle and breast before other feathers are moulted; the tail streamers are absent until the wholly black male plumage is attained.

**VOICE:** The song is an ascending whistled *"pli pli pli pli"* and the alarm call is a harsh *"zweet!"*.

**BEHAVIOUR:** Perches for long periods, then sallies to catch insects in flight, or to glean insects from leaves. Pairs and family parties chase each other excitedly and restlessly through the trees.

**BREEDING:** Breeds throughout the northwest monsoon (September–April), placing its woven cup-shaped nest, normally containing just one egg, on a thin hanging branch at the edge of a clearing in the woods.

## E Seychelles Bulbul

2

*Hypsipetes crassirostris* — Merl Sesel

Length: 23cm/9·1in Wingspan: 36cm/14in

A common endemic resident, frequenting all wooded habitats on the larger islands, but absent from all islands smaller than La Digue.

**IDENTIFICATION:** A noisy, inquisitive thrush-like bird, which is slim-bodied and relatively long-tailed with a rather upright perching posture; it is uniformly olive-brown with a large orange bill and a distinctive spiky crest. It imitates the calls of other species including Common Myna and Seychelles Black Parrot, but cannot be confused with either species if seen well. ADULT: The plumage is largely dull brownish-olive, greyer on the face and breast, and paler on the lower belly and vent. The crown is blacker and conceals a spiky crest, which is raised when excited **(2a)**. Russet edgings to the upperwing coverts are revealed in flight, as are the pale tip and edges to the longish, square-ended tail. The bill is relatively large and bright orange, the eye is dark brown, and the legs are bright orange. **JUV:** Similar to adult, but lacks the spiky crest, the russet flash in the wing is larger and more pronounced, the bill is dark, and the legs are duller and greyer.

**VOICE:** Has a very varied repertoire of nasal whistles and throaty cackles. Common calls are a repeated nasal *"tchaay"*, a wheezy *"weeez"* and a nasal series of *"chek"* notes, which may develop into a rapid crescendo. Calls are usually harsher and less melodious than the Common Myna, but both species are accomplished mimics.

**BEHAVIOUR:** Moves through trees in noisy groups, often involving a pair with up to two youngsters. It may lurk in dense foliage at times but also perches in the open, and may approach the observer inquisitively in denser forest.

**BREEDING:** Nests containing two eggs are built in the fork of a tree during the northwest monsoon (October–March), the fledged young remaining with their parents during the first half of the southeast monsoon (May–July).

1f

1m

2

2a

## E Seychelles Warbler 1

*Acrocephalus sechellensis* Timerl-de-Zil Sesel

Length: 14 cm/5·5 in Wingspan: 17 cm/6·7 in

Once a critically endangered endemic restricted to a small area of native woodland on Cousin, but habitat restoration and translocations have improved the outlook for this species. There are now thriving populations on Cousin (300), Cousine (200) and Aride (>2,000), where numbers continue to increase in regenerating native forest. Introduced to Denis in 2004.

CONSERVATION STATUS: **Vulnerable**

**IDENTIFICATION:** A small drab olive-brown warbler, lacking truly distinctive features other than its voice. It is slimmer, longer tailed and thinner-billed than the similarly drab female Seychelles Fody *(page 80)*. **ADULT:** The upperparts are uniform olive-brown and the underparts are paler, with grey-olive on the belly and flanks and a pale, grey-washed vent. There are diffuse grey and yellowish streaks on the breast, and the throat is a paler yellowish-white. There is a diffuse and incomplete buffy-yellow eye-stripe. The long, slim bill has an orangey base, the eyes are red-brown, and the longish legs are blue-grey. **JUV:** Much as adult, but the tail often appears shorter and the iris is slate-grey, turning brown with age.

**VOICE:** This species has a very varied repertoire, including harsh *"check"* alarm notes, a repeated *"pea-cher, pea-cher"*, a musical *"woddle-wee!"*, a complex melodious warbling song, and a quieter *"twee-twee-twee-twee-twee"* sub-song. Begging fledglings give an incessant grating *"witch, witch, witch..."*.

**BEHAVIOUR:** Gleans insects from the underside of leaves, making pronounced bill-clacking sounds. It also flycatches, and even forages on the ground. It is not shy and often feeds on open branches and may approach the observer inquisitively. Each pair establishes a well-defined territory, which is occupied year-round, the adults often being helped to raise their young by offspring from previous broods.

**BREEDING:** Nests year-round, although there is a distinct peak in nesting at the start of the southeast monsoon (April–June). A clutch of one or two eggs is laid in an open cup nest concealed high in the branches of a tree.

## E Seychelles Magpie-robin 2

*Copsychus sechellarum* Pisantez

Length: 22 cm/8·7 in Wingspan: 29 cm/11·4 in

An endangered endemic resident, now occurring naturally only on Frégate (70 birds). An ongoing recovery programme has involved its re-introduction to Cousin (25 birds), Cousine (25 birds) and Aride (20 birds), and further introductions are planned to other rat and cat-free islands.

CONSERVATION STATUS: **Endangered**

**IDENTIFICATION:** Unmistakable, a piebald robin, which is black with bold white wing-patches. The Common Myna *(page 84)* has similar wing-patches but is larger, browner, shorter-tailed and more starling-like in behaviour. **ADULT:** The plumage is entirely glossy black, with a bluish sheen when seen in good light, relieved only by bold rounded white patches on the wing coverts. These are visible as a broad white bar on the folded wing. The square-ended tail is relatively long and is often cocked when the bird alights. The black bill is relatively robust, and the longish legs and feet are black. The eye is dark. **JUV:** Similar to adult, but the white wing-bar is broken with diffuse black marks.

**VOICE:** Calls include a melodious whistled *"whee woo"*, ascending on the first syllable and descending on the second, a repeated ascending *"weeh weeh weeh"*, and a single ascending *"wooeee"*. The song, which is produced by both sexes, is a mournful descending series of whistled notes: *"weeh werh werh whoo"*. The sub-song includes a random series of whistles, whirrs and clicks.

**BEHAVIOUR:** Generally tame and inquisitive, this species boldly follows giant tortoises and humans in search of revealed prey. It hops on the ground, turning leaf-litter in the manner of a thrush, but also gleans insects from leaves, fly-catches and attacks geckos, skinks etc. Pairs defend their territory against intruders, often including their own fledged offspring from previous broods.

**BREEDING:** Nesting occurs year-round, a single egg usually being laid in a hole in a tree, coconut crown or nest-box provided for their use.

1

2

## 1 Madagascar Fody

2

*Foudia madagascariensis* Kardinal, Sren

Length: 14cm/5·5in Wingspan: 18cm/7·1in

A widespread and abundant species, occurring on all islands. It is most numerous in open lowland habitats with smaller numbers being found in the more densely forested uplands. The origin of Seychelles birds is debated, but it is assumed to have been introduced in the 1800s.

**IDENTIFICATION:** The bright red plumage of breeding males is unmistakable. Females and non-breeding birds are relatively nondescript fawn-brown finch-like birds. However, their dark-streaked upperparts and short conical bill separate them from the similar Seychelles Fody. **ADULT:** Males in breeding plumage **(1m)** are bright orangey-red above and below, relieved only by blackish streaks on the mantle, dark brown flight feathers with paler brown edges, and a mid-length dark brown tail. There is a small black area around the dark eye, the legs are pink and the short conical bill is blackish. Males moulting in and out of breeding plumage often show a grey or sometimes yellowish belly and lower breast; on the larger islands they often breed in this plumage. Non-breeding male and females **(1f)** are olive-grey-brown above with dark streaking. The underparts are a paler grey-brown and unstreaked. There is a pale buffy line over the eye and a buffy crescent below, imparting an almost 'spectacled' look. The bill and legs are pinkish-grey. **JUV:** Similar to non-breeding adult, but shows a warmer yellowish tone to the underparts. A yellowish gape is obvious in recently fledged birds.

**VOICE:** A very vocal species, often emitting a single high-pitched *"sip"*, a louder chirruping *"tsiiiip"* or a rapid trilling series of *"sip sip sip sip sip"* notes from a perch. Males occupy breeding territories and proclaim them with a high-pitched hissing song.

**BEHAVIOUR:** Highly commensal, feeding on the ground around houses and in open grassy areas. It forms larger flocks outside the breeding season. Fledglings beg with incessant strident calls and rapid shaking of their wings.

**BREEDING:** Breeding occurs throughout the year, but most birds nest during the northwest monsoon (October–March), building domed nests of grasses suspended high in the branches or tucked in a coconut crown, and laying two or three pale blue eggs.

## E Seychelles Fody

1

*Foudia sechellarum* Toktok

Length: 12cm/4·7in Wingspan: 17cm/6·7in

A localised endemic, restricted to Frégate, Cousin and Cousine Islands, with a total population of *ca.* 3,000 birds. In 2002, birds were translocated successfully to Aride, where it was formerly a rare vagrant, and in 2004 to Denis.

CONSERVATION STATUS: **Vulnerable**

**IDENTIFICATION:** A dull olive-grey-brown finch-like bird. It is similar to Madagascar Fody in non-breeding plumage but is more uniform and lacks streaking on the upperparts. It also appears larger and stockier, with a shorter tail, shorter legs and a longer, slightly down-curved bill. **ADULT:** Males in breeding plumage have a yellow fore-crown and throat, but are otherwise uniformly olive-grey-brown with slightly paler underparts and a dark line through the eye. The black bill is long and slightly down-turned. The short, greyish legs and short tail impart a dumpy look. Non-breeding males and females lack the yellow throat and fore-crown. **JUV:** Similar to non-breeding adult.

**VOICE:** A very vocal species, feeding groups constantly emitting scratchy *"tsk tsk"* whistles, and louder *"tok tok"* calls. The song of the male consists of a series of *"tweet"* notes.

**BEHAVIOUR:** Feeds on the ground around houses in small groups where it is often bold, dominating the smaller Madagascar Fody. It also feeds on insects and seeds high in trees, when its presence is revealed by the constant calls. Some individuals habitually feed on seabird eggs.

**BREEDING:** Breeding occurs throughout the year, with a peak at the end of the southeast monsoon (September–October). The nests are untidy balls of grasses, often placed low in bushes, but also higher in trees, and contain two or three white eggs.

1m
1f

2

## E Seychelles White-eye 1

*Zosterops modesta* Zwazo Linet Sesel

Length: 10 cm/3·9 in Wingspan: 15 cm/5·9 in

Severely endangered remnant population in mid-elevation secondary growth with alien trees and shrubs on Mahé (<50 birds), but thriving population on tiny offshore island of Conception (*ca.* 300 birds), discovered only as recently as 1997, from where 31 birds were translocated successfully to Frégate in 2001.

CONSERVATION STATUS: **Critically Endangered**

**IDENTIFICATION:** A very small dull olive-grey bird, easily overlooked amongst the more abundant and noisier Seychelles Sunbirds, but from which it differs in having a tiny pointed bill and a clear white ring around the eye. **ADULT:** The plumage is uniformly dark grey-olive above and paler grey below and on the lores, palest on vent. The clear white eye-ring is the main identification feature. The tail and wings are short, as is the sharp black bill, and the legs and eye are black. **JUV:** Similar to adult although the eye-ring is less distinct.

**VOICE:** Foraging birds call a repeated diagnostic trilling *"chee-err"*, as well as soft clipped *"tik"* contact notes.

**BEHAVIOUR:** Feeds in small groups, flitting about nervously in trees and bushes. It flies short distances between trees with a rapid whirring flight on short wings.

**BREEDING:** Breeds mostly during the northwest monsoon (October–March), the nest being a woven cup located in a dense shrub or tree in which two pale green-blue eggs are laid.

## E Seychelles Sunbird 2

*Nectarinia dussumieri* Kolibri Sesel

Length: 12 cm/4·7 in Wingspan: 16 cm/6·3 in

A widespread and common endemic, frequenting wooded habitats on all but the very smallest islands.

**IDENTIFICATION:** A very small restless species, which emits frequent calls as it forages for nectar and insects in flowering trees and shrubs. It appears largely dark olive-grey, paler below, with a short tail and a long decurved bill. The adult males are unmistakable, having a metallic blue breast and orange pectoral tufts. **ADULT:** Males **(2m)** are uniformly dark olive-grey above, darker still on the flight feathers, with a metallic blue-black breast and throat, and a paler grey belly and vent. There are bright orange pectoral tufts on either side of the breast, which are prominently revealed during display but are otherwise normally concealed by the closed wing. The short tail has broad whitish tips on the underside, which are prominent in flight. The blackish bill is long, slim and decurved, and the eye and legs are black. Females **(2f)** are similar to males but lack both the metallic-blue breast and the pectoral tufts, the breast and throat instead being mid-olive-grey, which merges with the paler grey belly and vent. **JUV:** Much as female, although young birds often show a yellowish gape. Young males **(2y)** may show black speckling on the breast as they attain adult plumage.

**VOICE:** Foraging birds frequently emit a penetrating *"tsiuu"*. The alarm calls include *"wit"*, *"wiii"*, and a more urgent harsh rolling *"wrrrit"*. The song, which is given by both males and females, is a repetitive jumble of wheezy whistled notes.

**BEHAVIOUR:** An active, restless and often noisy species, which forages with quick jerky movements for nectar and insects in flowering trees and shrubs and in fruiting trees. Males hold territory throughout the year and gather towards dusk to lek in a favoured tree. This often results in a noisy assemblage of dozens of displaying males, which attracts the attention of spectating females.

**BREEDING:** Breeding occurs throughout the year, although more frequently in the northwest monsoon (October–March). The nest **(2ne)**, in which a single egg is laid, is an enclosed woven ball with an entrance near the top that hangs from long strands often high in a tree.

1
1
2ne
2f
2y
2m

## 1 Common Waxbill

1

*Estrilda astrild* — Bengali

Length: 11 cm/4·3 in — Wingspan: 13 cm/5·1 in

A presumed introduced species, which favours moist areas of rank seeding grass. It occurs only on La Digue, where it is common, and on Mahé, where it is only locally common along the west coast.

**IDENTIFICATION:** A very small, long-tailed, short-winged grey-brown finch, which feeds inconspicuously before flying on, often as a group, and giving distinctive calls. Close views reveal the diagnostic waxy red bill, a red area around the eye, and a reddish patch on the lower belly. **Adult:** The upperparts are grey-brown and the underparts are paler grey-brown faintly barred darker on the breast, flanks, wings and tail. There is a reddish-pink stripe on the lower belly, and the vent and undertail are black. The tail is relatively long and slightly graduated towards the tip. The beady black eye is surrounded by a bright red patch, the stubby bill is bright red, and the tiny legs are black. **Juv:** Similar to adult but lacks the red patches around the eye and on the belly, and the bill is brownish-black. **VOICE:** A distinctive nervous *"ping, ping"* call is given by flocks in flight, and feeding birds give a chattering *"cher, cher, cher"*. **BEHAVIOUR:** Feeds on the ground or climbs grass stems to feed on the seed-heads. It is normally seen in nervous twittering groups of up to 30 or so birds, but is rather alert and difficult to approach closely. **BREEDING:** Breeding occurs during the northwest monsoon (October–March), clutches of 4–6 eggs being laid in domed nests of grass built in low bushes.

## 1 Common Myna

2

*Acridotheres tristis* — Marten Ordiner

Length: 25 cm/9·8 in — Wingspan: 35 cm/14 in

An abundant introduced species, occurring almost everywhere, but particularly numerous in more open man-made habitats.

**IDENTIFICATION:** A conspicuous chunky starling, mostly blackish with bold white wing-patches, a white tip to the tail in flight, and a yellow bill and bold yellow patch around the eye. **Adult:** The head and upper breast are blackish and the mantle, wings and uppertail coverts are brownish-black. Bold rounded white patches in the wing contrast with the black primaries in flight, and the slightly graduated mid-length blackish tail bold has a white tip. The underparts are brown with a white vent. The bill is thickset and pointed and orange-yellow, and there is a bright yellow patch below the dark eye. The legs are sturdy and bright orangey-yellow. Some birds lose their head feathers, revealing a strikingly yellow head. **Juv:** similar to adult although the yellow facial skin is less distinct and the head is browner. **VOICE:** A noisy species, with a wide range of loud and shrill whistled calls, grating notes, chuckles and wheezes, which reach a crescendo in evening roosts. Typical calls include a clear ringing *"piuu"*, a whistled *"tew tew"*, and an abrupt *"chup"*. It readily mimics other species, including Seychelles Bulbul and Seychelles Black Parrot. **BEHAVIOUR:** Commonly feeds on the ground, but is often rather wary. The flight is strong, with rapid beats of the stubby rounded wings. **BREEDING:** Breeds throughout the year, laying two or three eggs in a scruffy bundle of grass placed in a coconut palm or the roof of a building.

## 1 Feral Pigeon

3

*Columba livia* — Pizon Domestik

Length: 33 cm/13 in — Wingspan: 65 cm/26 in

An introduced species restricted to open and urbanised areas on the larger islands, and only a casual visitor elsewhere.

**IDENTIFICATION:** A familiar bird, with a bewildering variety of plumages that range from all-white to largely dark grey. It is only likely to be confused with Madagascar Turtle Dove *(page 70)*, which is smaller, squatter, mostly vinous-chestnut with a paler grey head, and more tied to forested habitats. **Adult:** The typical plumage is bluish-grey, darker grey on the head and breast, with two black wing-bars and a white rump, and often with a metallic green sheen on the neck. It also occurs in an all white form, variegated white and grey, all dark grey, or even a piebald white and rufous brown. Flocks often comprise a mix of these forms and intermediates. The bill is small and normally grey with a fleshy base, and the legs and feet are flesh-pink. **Juv:** Similar to adult. **VOICE:** Gives a range of dove-like *"coo"* calls. **BEHAVIOUR:** Feeds on the ground, often on pavements, in gutters etc. in small groups. **BREEDING:** Breeds year-round and nests on ledges. The clutch normally comprises two white eggs and the chicks are fed on crop 'milk' produced by the parents.

1

2

3

# LAND MAMMALS

Lying so far from other land-masses, and having become isolated as islands before many mammals had evolved, it is not surprising that the land mammal fauna of Seychelles is depauperate. There are only two native land mammals, both of which are bats and are endemic. The Seychelles Fruit Bat remains common and widespread, but the Sheath-tailed Bat is poorly known and possibly severely endangered. Commensals of man were quick to colonise the islands and have no doubt had important but largely unknown effects on the native plants and animals. These include a tenrec, two species of rat, one mouse, a rabbit and a hare. Other species, not included here, have become feral, including domestic cats and dogs, and there were formerly feral populations of goats on islands such as Aride, which have since been removed.

## MEGACHIROPTERA

### E Seychelles Fruit Bat 1

*Pteropus seychellensis* Sousouri

Length: *ca.* 25 cm/10 in Wingspan: 100 cm/39 in
Weight: 540 g

A widespread endemic resident, common in wooded habitats on all the larger islands, travelling to and from smaller islands according to the availability of fruit. It is taken legally in some numbers for human consumption.

**IDENTIFICATION:** Unmistakable; a very large bat, which flies with slow, deliberate beats of slightly-bowed, broad wings. May recall a large bird of prey, but there are none of this size in Seychelles. **Adult**: Furry; bright yellow-orange head, upper mantle and underparts contrast with blackish wings, lower back, rump, and muzzle. The underparts are often darker towards the blackish legs; some individuals have largely black underparts with a rufous band across the chest. The eyes are large and dark and the ears are mid-sized, pointed and blackish. **Juv**: Much as adult, but smaller and the body colour is less bright. **VOICE:** Extremely noisy when feeding in groups at night or when squabbling at a day-time roost, producing a range of harsh chatters, piercing shrieks and wing claps. **BEHAVIOUR:** Feeds exclusively on fruits and nectar; most active at night, although often flies during day, particularly in the early morning and at dusk. Flies high over mountains or when making sea-crossings. Hangs upside-down from branches high in communal roosting trees during the day **(1a)**. Sadly, electrocuted bats are a common sight on roadside wires. **BREEDING:** Females give birth in November and December, and the family occupies a small territory in the roost. Juvenile bats become independent by April and often occupy the centre of roosts at the start of the southeast monsoon.

## MICROCHIROPTERA

### E Sheath-tailed Bat 2

*Coleura seychellensis* Sousouri Bannann

Forearm length: male 5·4 cm/2·1 in; female 5·6 cm/2·2 in
Weight: 10·2–12·2 g
(males average 10·2 g; females average 11·1 g)

A very rare and probably highly threatened endemic species. Recorded from Praslin, La Digue, Mahé and Silhouette in the 1990s, but known active roosts are currently restricted to Silhouette. Roosts in boulder-caves and feeds under the canopy in moist forests.

CONSERVATION STATUS: **Critically Endangered**

**IDENTIFICATION:** The only small insectivorous bat in Seychelles. As far as is known this bat is strictly nocturnal, flying low under the canopy. **DESCRIPTION:** A typical very small, insectivorous bat, with a dark brown furry body, membranous wings, and a very short tail projecting from the upper-surface of the membrane extending between the hind legs. It has flaps of bare skin on the face and mid-sized pointed ears. **VOICE:** Emits high-pitched cries when disturbed at their roost, and a high-pitched stridulating sound when feeding. **BEHAVIOUR:** Clings to the undersides of roofs or sides of boulder-caves when roosting during the day. Rarely observed feeding. Nocturnal, flying back and forth along a set route, often quite low to the ground, veering occasionally from side to side. Feeds on flying insects > 0·5 cm in length which it finds by means of echolocation. **BREEDING:** Poorly known.

1

1

1a

2

RODENTIA

## I Ship (Black) Rat

1

*Rattus rattus* Lera

Length (including tail): 35-48 cm/14–19 in

An introduced species, common and widespread on all but a handful of islands, where it has a largely unknown but probably important impact on native species. Found from coasts to tops of high mountains, but apparently cannot co-exist over an extended period with the larger Norway Rat.

**IDENTIFICATION:** A mid-sized to large, furry rodent with a long, sparsely-furred tail; smaller and sleeker than Norway Rat, with proportionately larger ears. Individuals in Seychelles tend to be smaller than elsewhere, with a relatively longer tail.

**DESCRIPTION:** The colour of the upperparts is immensely variable, ranging from grey to brown (rarely black), whilst the underparts can be white, intermediate in colour or as dark as the back. The long, pinkish sparsely-furred tail may aid their often arboreal habit. The ears are relatively large. **VOICE:** Emits high-pitched squeaks.

**BEHAVIOUR:** Very catholic in its choice of habitat and diet, foraging opportunistically on available prey. Climbs nimbly, and equally at home in trees as on the ground; also clambers along wires to gain access to buildings.

**BREEDING:** Can breed throughout the year, but probably less young are produced during the drier southeast monsoon (June–September).

## I Norway (Brown) Rat

2

*Rattus norvegicus* Lera

Length (including tail): 38–50 cm/15–20 in

Introduced species probably occurring together (at least at times) with Ship Rat on Mahé, Praslin and La Digue, with single-species populations restricted to a few offshore islands such as Conception and Frégate (the latter extirpated in 2000), but with the potential to turn up anywhere.

**IDENTIFICATION:** A large, furry rodent with a long, sparsely-furred tail. Adults are much larger and bulkier than Ship Rats, with proportionately smaller ears.

**DESCRIPTION:** The colour of the upperparts is generally mid-brown or grey-brown; the underparts are paler brown.

**VOICE:** Emits high-pitched squeaks.

**BEHAVIOUR:** Very catholic in its choice of habitat and diet, foraging opportunistically on available prey. Its relatively large size dictates a more terrestrial life-style than the Ship Rat.

**BREEDING:** Can breed throughout the year, but probably less young are produced during the drier southeast monsoon (June–September).

## I House Mouse

3

*Mus domesticus* Souris

Length (including tail): 15–20 cm/6–8 in

An introduced species which is widespread in Seychelles, being present on all but a handful of the offshore islands. It is generally scarce away from human habitation, but is more generally distributed on Aride and Frégate.

**IDENTIFICATION:** The familiar small furry rodent with a mid-length sparsely-furred tail. Much smaller than Ship Rat, with a proportionately shorter tail and larger eyes.

**DESCRIPTION:** The fur colour is uniformly grey-brown, paler below, and with little variation between individuals. The sparsely-furred longish tail is pinkish-grey. The eyes are large and blackish.

**VOICE:** Emits high-pitched squeaks.

**BEHAVIOUR:** Largely nocturnal, but also active during dull weather. It is a generalist feeder, taking whatever is abundant at the time. It is an agile climber, feeding in trees as well as on the ground and inhabiting buildings.

**BREEDING:** Breeds throughout the year, with more young being produced during wetter periods.

1
2
3

INSECTIVORA

## 1 Common Tenrec

1

*Tenrec ecaudatus* Tang

Length: 28 cm/11 in

An introduced species, confined to, and common on, Mahé and Praslin.
Scampers about noisily in the undergrowth of forested areas up to the cloud forest.

**IDENTIFICATION:** Unmistakable: a terrestrial, hedgehog-like animal.
**DESCRIPTION:** An elongated, hedgehog-like animal, covered in long, light-brown hair interspersed with spines particularly around the neck. The face and long snout are paler and pinker and lack the spines. The legs are quite slim, and there is no obvious tail. The ears are mid-sized and pinkish. There are long, sensitive whiskers around the face, and the eyes are small. The jaws are powerful, with large, canine teeth.
**VOICE:** Generally silent, but produces characteristic rustling sounds as it snuffles about in the leaf-litter.
**BEHAVIOUR:** Trundles actively and quickly through the undergrowth of woods in search of fruits and invertebrate prey. Males and immatures are active throughout the day, though females are more nocturnal. Its eyesight is poor and animals often stop to sniff the air; they can be very confiding.
**BREEDING:** Aestivates in burrows during drier periods of weather. Most young are born during the first part of the wetter northwest monsoon (October–December).

LAGOMORPHA

## 1 European Rabbit

2

*Oryctolagus cuniculus* Lepan

Length: 45 cm/18 in

An introduced species, which is numerous only on the small islands of Recif, Mamelles and Chauve-Souris (Cote D'Or Praslin), but commonly escapes from captivity on larger islands, particularly Mahé. It was eradicated from Bird in 1995.

**IDENTIFICATION:** A very familiar mammal, squat in shape, with long ears, large eyes, long back-legs and short 'button-tail'. Does not occur on Cousin, the only island in Seychelles where the Black-naped Hare occurs.
**DESCRIPTION:** The fur is extremely variable in colour, ranging from white, through brown or grey, to black. On Recif, black animals dominate. Rabbits in Seychelles tend to be larger and more sluggish than 'wild' rabbits, with longer, less upright ears.
**VOICE:** Generally silent.
**BEHAVIOUR:** Active throughout day, but tends to be crepuscular, grazing on grass and short herb swards, and bounding into denser cover when threatened.
**BREEDING:** Poorly known in Seychelles.

## 1 Black-naped (Indian) Hare

3

*Lepus nigricollis* Lyev

Length: 55 cm/22 in

Native to southern India and Sri Lanka, introduced to Cousin, via Mauritius and Java, in the 1920s–1930s, where it remains in small numbers. It is poorly known because of its largely nocturnal habits.

**IDENTIFICATION:** A thin, scrawny hare, larger and slimmer than European Rabbit and with relatively longer ears.
**DESCRIPTION:** The fur is a pale grey-brown, paler below, and with a marked black patch on the nape. The ears are mid-length and held erect.
**BEHAVIOUR:** Largely nocturnal, but some individuals feed around the houses on Cousin prior to dusk. Grazes on swards of grass and herbs.
**BREEDING:** Poorly known in Seychelles.

1

2

3

## MARINE MAMMALS

Marine mammals remain poorly known in Seychelles waters and much awaits discovery, particularly in the deeper waters off the Seychelles Bank. Only the Bottlenose Dolphin is common in the inshore waters around the islands, and this is the only species likely to be seen by the largely land-based observer. Boat trips taken to the more distant islands of Bird and Denis, and especially to the deeper water beyond these islands, may reveal a wider array of species, of which the four most frequently seen are included below. Owing to their mid-oceanic position, the waters around Seychelles are likely to host many other species at different times of the year, and all sightings should be carefully documented and animals photographed wherever possible. Prior to human occupation, it is probable that Dugong *Dugong dugon* would have grazed the sea-grass beds of the islands' lagoons, whilst early records hint at the presence of a large species of seal (possibly a fur-seal *Arctocephalus* sp.) on Bird Island. Both species would no doubt have succumbed quickly to exploitation for food.

### CETACEA

### Sperm Whale (1)

*Physeter macrocephalus*

Length: 18m/60ft

Hunted historically in deeper waters off Bird and Denis Islands, this species is still relatively common in deeper waters mainly south and west of the Seychelles Bank.

**IDENTIFICATION:** The only common large whale in deeper waters around the northern Seychelles, though unlikely to be seen from shore.

**DESCRIPTION:** The massive forehead is distinctive, with the blow-hole positioned at the tip and slightly to the left, producing a distinct, angled 'blow'. Very dark grey in colour, with paler areas on the belly and around the mouth. The skin appears wrinkled, with a small hump and smaller ridges towards the tail. The jaw is very long and narrow, with conical teeth. The tail flukes are broad and triangular with a straight trailing-edge.

**BEHAVIOUR:** Can occur singly or in large pods. Spends long periods at the surface prior to lengthy and often deep dives, with the large, upward-sloping forehead showing above the water.

**BREEDING:** Adults are often accompanied by calves, suggesting that the area may be a breeding and calving ground.

### Short-finned Pilot Whale (2)

*Globicephala macrorhynchus*

Length: 5m/16ft

A small, dark, toothed whale, which occurs in both inshore and offshore waters, but is more numerous offshore.

**IDENTIFICATION:** A small whale, separated from the larger dolphins by its very dark colouring, narrow body with bulbous head and no obvious beak. The dorsal fin has a hooked tip. The explosive blow is distinctive.

**DESCRIPTION:** The body is uniformly blackish, with a paler, greyish 'saddle' on the back behind the dorsal fin, and a white patch on the belly. The small, slim shape is distinctive, and the rounded head lacks an obvious beak. The dorsal fin is positioned well forward towards the head.

**BEHAVIOUR:** Occurs in both small and large pods. It is scarce in inshore waters, pods occasionally being observed between islands, but is more common offshore. This species is not particularly inquisitive.

**BREEDING:** Poorly known.

1
1

2

## Spinner Dolphin ❶

*Stenella longirostris*

Length: 2 m/7 ft

An occasional visitor to inshore waters around the granitic islands.

**IDENTIFICATION:** A small, slim dolphin separated from other species by its relatively long and pointed fins, gently-sloping forehead, very long beak and acrobatic habits.
**DESCRIPTION:** Distinctive two-toned appearance, dark grey above and paler grey/white below, with a pronounced blackish line from the tip of the beak through the eye and to the front flippers.
**BEHAVIOUR:** Occurs in groups of variable size. Very acrobatic, regularly leaping clear of the water and spinning distinctively in mid-air; often rides the bow-waves of boats.
**BREEDING:** Poorly known in Seychelles.

## Bottlenose Dolphin ❷

*Tursiops truncatus* Marswen

Length: 4 m/13 ft

The common dolphin of inshore waters in Seychelles, with resident pods around all islands, often seen from the shore. Harpooned for food as recently as the 1970s, the recovering population is still relatively wary of man.

**IDENTIFICATION:** A large, grey dolphin with relatively short fins, a steep forehead, and a short beak.
**DESCRIPTION:** The largest dolphin, dark grey above, with pinkish-grey sides and a whitish belly, lacking strong contrasts in body colour. The body is long and chunky, and the fins and tail flukes are relatively small and pointed. The forehead is steep, meeting the short thick beak in a distinct crease.
**BEHAVIOUR:** Occurs in small pods of up to 15 animals, which appear to hold territories year-round. It is inquisitive and may approach boats, but the population in Seychelles is generally more cautious than elsewhere. It is often acrobatic, leaping clear of the surface, and slaps its tail on the surface and surfs on breakers.
**BREEDING:** Young calves appear in November, at the start of the calm and warm northwest monsoon.

## Risso's Dolphin ❸

*Grampus griseus*

Length: 4 m/13 ft

A rare species, occasionally seen in offshore waters.

**IDENTIFICATION:** A distinctive large, pale grey dolphin with a very tall, slightly curved dorsal fin, and steeply-domed head lacking an obvious beak.
**DESCRIPTION:** The whitish-grey body is covered in conspicuous scratches and paler markings, and the flippers and tail flukes are darker. The blunt head lacks a beak and bears a 'V'-shaped furrow on the forehead. The flippers are very long and slender, and the tail flukes are curved and serrated.
**BEHAVIOUR:** Occurs in small groups, which can breach acrobatically.
**BREEDING:** Poorly known in Seychelles.

1

2

3

## REPTILES

Seychelles is blessed with a wide-range of reptiles. A total of 22 species of reptile are currently recognised as being native to the islands (excluding the giant tortoises), these include four marine turtles (two of which breed in Seychelles), three terrapins, nine geckos, four skinks, one chameleon, and three snakes, together with occasional records of the Yellow-bellied Sea Snake *Pelamis platurus*. There are also reports of a feral population of an escaped terrapin species, the Red-eared Slider *Trachemys scripta* on Mahé, and of an invasive lizard species *Calotes versicolor* on St. Anne and Mahé. Estuarine Crocodiles *Crocodylus porosus* were present in the mangrove-lined bays prior to human settlement, but were quickly hunted out. Place names such as Roche Caiman bear mute testament to their passing.

### LAND TORTOISES (PAGE 98)

Although giant tortoises were once common on many of the islands of Seychelles, the only wild population remaining is that at Aldabra. Feral populations of giant tortoises reintroduced from Aldabra occur on several islands including Curieuse, Frégate, Cousin, Cousine, and North, and these give a flavour of a more ancient time. In addition, specimens believed by some people to bear features of two previously described species of giant tortoise are being reared in captivity on Silhouette by the Nature Protection Trust of Seychelles (NPTS).

### TURTLES & TERRAPINS (PAGES 100–103)

The two species of marine turtles that breed in Seychelles, were so vulnerable on the nesting beach that their numbers were greatly diminished by over-harvesting during the past three centuries. Although the nesting populations are currently much smaller than their historical levels, enforcement of protective legislation has resulted in significant recovery at several nature reserves. Nevertheless, Green Turtles remain rare in the inner islands of Seychelles where they are heavily poached. Two other turtles, the Loggerhead *Caretta caretta* and the Leatherback *Dermochelys cariacea* feed within Seychelles waters, but are rarely seen. Habitat loss is the greatest threat to the native terrapin species, one of which is probably already extinct. The remaining two can still be seen with diligence and luck, at wetland sites on La Digue, Praslin, Mahé and Frégate. Both species are the subject of a captive-breeding programme conducted by the NPTS on Silhouette.

### GECKOS (PAGES 104–109)

There are nine currently recognised species of gecko include seven endemic species and two widespread commensals. Many, such as the day geckos are brightly-coloured and conspicuous, whilst others are more cryptic in their colouration. One commensal species, the Pacific Gecko, is abundant inside houses throughout the islands. The bright green day geckos and the drabber bronze gecko group pose can pose identification problems, but careful observation of the salient features, together with location and habitat, should enable an observer to put the correct name to an individual. The day geckos are widespread in many types of vegetation whereas the bronze geckos are more confined to areas of palm forest or rat-free islands.

### SKINKS & CHAMELEON (PAGES 110–115)

All four skink species and the single chameleon are endemic to Seychelles. The Seychelles Skink is widespread and easily seen, whilst the Wright's Skink is abundant on a few rat-free islands where it can reach impressive densities. Skinks on St. Anne and Mahé should be checked thoroughly as both islands are home to an apparently increasing population of the introduced lizard *Calotes versicolor*. The remaining skinks are cryptic species, which require much searching for in appropriate habitat. The chameleon can be seen with luck by diligent observers searching path-side vegetation for its slow awkward movements.

### SNAKES (PAGE 116)

The three resident snake species are rarely seen. The wolf snake is diurnal and can be encountered foraging across the forest floor or climbing into trees in search of geckos. The house snake is nocturnal and is most often encountered when it enters houses in search of prey. The tiny blind snake lives an even more cryptic subterranean life, but may be forced to the surface during periods of heavy rain. All of the resident snakes are harmless.

## Aldabra Giant Tortoise

1

*Dipsochelys dussumieri* Torti D'ter

Length: up to 1·4 m/5 ft  Weight: up to 250 kg

A well-known species, endemic to Aldabra but commonly kept in captivity throughout the granitic Seychelles with feral populations on several islands. The original endemic giant tortoise species from the granitic islands are long extinct in the wild, but individuals retaining certain characteristics of two of these species (Seychelles Giant Tortoise and Arnold's Tortoise) are the subject of a captive breeding programme on Silhouette (see below).

**IDENTIFICATION:** Unmistakable: a huge, slow-moving tortoise. The only free-ranging tortoise species in Seychelles. Individuals may live for over 100 years.

**DESCRIPTION:** A very large tortoise with a high-domed, dark grey-brown carapace and very thick, wrinkled greyish skin. Males are larger than females, with a longer tail and a depression on their underside. The front legs are club-shaped and the back legs are thick and sturdy in order to support the considerable weight of the body (up to 250 kg). The head is large and blunt, on a long, thick retractable neck. The eyes are dark and the nostrils are large and uppermost on the head.

**VOICE:** When frightened tortoises quickly pull their heads into their shell, making a hissing sound as they expel air from their lungs. Males roar loudly during copulation; females are silent.

**BEHAVIOUR:** Mostly active during the cooler parts of the day, particularly early in the morning and late in the afternoon, seeking shade and resting during the heat of the day. They are entirely herbivorous. Aldabra Giant Land Tortoises can swim well but only walk slowly on land. Individuals rarely interact other than during mating.

**BREEDING:** Mating normally occurs towards the end of the northwest monsoon (February–May), with egg-laying occurring during the drier southeast monsoon (June–September). Females lay 10–25 eggs in a nest-hole excavated at night and covered with sand. The eggs take 70–160 days to hatch depending on the temperature and the 70 mm long hatchlings emerge at night.

## Seychelles Giant Tortoise

2

*Dipsochelys hololissa*

## Arnold's Tortoise

3

*Dipsochelys arnoldi*

Torti D'ter

Length: up to 1·4 m/5 ft

The tortoises of the granitic islands are long extinct in the wild. Identification of tortoises is often contentious and genetic data conflict. However, individuals retaining certain characteristics of Seychelles Giant Tortoise *D. hololissa* and Arnold's Tortoise *D. arnoldi*, were located in various collections in the 1990s. The best representatives of both of these have been purchased and housed in separate enclosures on Silhouette by the Nature Protection Trust of Seychelles, where they are the subjects of a captive breeding programme.

**IDENTIFICATION:** Similar to the Aldabra Giant Land Tortoise, though the characteristics of these two species are as follows:

**Seychelles Giant Tortoise**: The carapace is elongated and bears numerous bumps, the channels between the plates across the back being flat and not curved convexly.

**Arnold's Tortoise**: The carapace is elongated, wider at the rear end with a very flat top, and bears a bold depression near the front on each side. The front entrance to the carapace is high above the neck; the front legs are long and the depressions in the carapace indicate strong muscle attachments. These features allow the tortoise to graze on higher vegetation whilst standing up.

1

2

3

TURTLES

## Hawksbill Turtle

1

*Eretmochelys imbricata* Kare

Length: up to 1 m/3 ft Weight: up to 60–70 kg

A widespread species frequenting shallow reef areas around all the islands. Heavily exploited for its shell in the past, populations are slowly recovering following the introduction of protective legislation enforced by the Seychelles Government. Small numbers still breed on the main islands, with larger rookeries on offshore and reserve islands.

CONSERVATION STATUS: **Critically Endangered**

**IDENTIFICATION:** A medium-sized sea turtle best separated from the similar Green Turtle by the large overlapping scales on its carapace, a distinctly protruding hooked upper beak and a relatively narrower, more pointed head.
**DESCRIPTION:** A large turtle with obvious overlapping scales on a rounded carapace. The carapace **(1ca)** is warm brown above, with patterns of black, reddish-brown and yellow. The underside is orange-yellow. The dark coloured head and body scales are typically separated by light coloured yellow or white borders. The hooked upper beak is distinctive. Hatchlings are rather uniformly dark brown in colour, although a little blacker on the underside **(1j)**.
**BEHAVIOUR:** Forages on sponges, soft bodied invertebrates, and algae on reefs. They are often very approachable at sea but very nervous on land when emerging onto the beach to lay eggs. They sleep at night in coral caves and under overhangs in the reef.
**BREEDING:** Hawksbills of the western Indian Ocean are unique in that they nest primarily in the daytime, while those elsewhere nest almost exclusively at night. They nest on sandy beaches mostly during the northwest monsoon (September–March). Females lay an average of about 170 eggs, but up to 250 eggs have been recorded, in a nest-hole dug on the beach platform, usually amongst beach-crest vegetation, in more shaded areas than the Green Turtle. Eggs take around 55–70 days to hatch, and hatchlings usually emerge *en masse* often at dusk to race to the sea. Females lay some 3–5 egg clutches in a season, but generally do not return to nest during two successive seasons. In Seychelles, they typically return to breed after intervals of two or three years.

## Green Turtle

2

*Chelonia mydas* Torti Dmer

Length: up to 1·5 m/5 ft Weight: up to 200 kg

Only tiny remnants of nesting populations remain in the inner islands of Seychelles where they have been heavily hunted for meat. More abundant in the outer islands, especially at the Aldabra group.

CONSERVATION STATUS: **Endangered**

**IDENTIFICATION:** A large sea turtle, best separated from the relatively smaller Hawksbill Turtle by the shape and size of its head, which is larger, bullet-shaped, and lacks a pointed beak. Unlike the Hawksbill, its carapace scales do not overlap.
**DESCRIPTION:** A large turtle with large dark greenish-brown or almost blackish plates on the carapace **(2ca)** that do not overlap. The carapace is obviously longer than wide. The underside is variably whitish or yellow. The upper and lower jaws are of equal length, lacking an obvious hook. The scales on the head scales are dark, ranging from blackish to red-brown, and often have a white edge. Hatchlings are very dark blackish-green above and white below, with bold white edges to the dark flippers and pale eyelids.
**BEHAVIOUR:** Forages on algae and sea grass in shallow inshore areas. It is generally not approachable at sea in Seychelles and is also very nervous on land, nesting only at night.
**BREEDING:** Females emerge to nest on sandy beaches during any month of the year, but especially during the months of February to April and June to September. Each female lays 150–200 eggs in a nest-hole dug in open areas at the top of the beach, always making much larger holes than Hawksbill Turtle. The eggs take around 55–70 days to hatch, and hatchlings usually emerge at night.

1 1j 1ca 2 2ca

## e Seychelles Black Mud Turtle

1

(Star-bellied Terrapin, Lesser Black Mud Turtle)

*Pelusios subniger parietalis* Torti Soupap

Length: up to 17 cm/7 in

Widespread in lowland freshwater wetlands, but overall numbers very low. Favours shallower, more vegetated swamps with still water. The subspecies *P. s. parietalis* is endemic to Seychelles.

CONSERVATION STATUS: **Critically Endangered**

**IDENTIFICATION:** The smallest Seychelles terrapin. Similar to Seychelles Yellow-bellied Mud Turtle, but has brown patches on the yellow underside, grey not yellowish skin on the neck and legs, and has distinctive bands of small scales leading from behind each eye to the nape, separating the larger plates on the head.
**DESCRIPTION:** Very small turtle with domed yellowish-brown carapace. The underside is yellowish with brown patches and the skin is grey. The top of the head is dark grey, sometimes with dark spots, and always lacks the wavy yellow and brown lines of the other two species.
**BREEDING:** Females lay 6–12 eggs, in 1–2 clutches, mainly during the wetter northwest monsoon (December–March), nesting in banks near marshes.

## e Seychelles Yellow-bellied Mud Turtle

2

(Yellow-bellied Terrapin, Chestnut-brown Mud Turtle)

*Pelusios castanoides intergularis*

Torti Soupap

Length: up to 22 cm/9 in

Widespread in deeper, more open lowland freshwater wetlands with flowing water, but overall numbers very low. The subspecies *P. c. intergularis* is endemic to Seychelles.

CONSERVATION STATUS: **Critically Endangered**

**IDENTIFICATION:** Very similar to the Seychelles Black Mud Turtle but differs in its primarily yellow, not black, underside (plastron). Adults are larger and lack a band of small scales leading from behind each eye to the nape. It has yellowish skin (not grey) on the neck and legs.
**DESCRIPTION:** A small turtle with a domed, rather elongated yellowish-brown carapace bearing darker brown blotches. The underside is plain yellow and the skin is yellowish, except for the top of the head, which is covered in wavy brown and yellow lines. The head is composed mostly of larger plates, smaller scales being confined to the nape.
**BREEDING:** Females lay 6–12 eggs, in 1–2 clutches, mainly during the wetter northwest monsoon (December–March), nesting in banks near streams.

## E Seychelles Mud Turtle

(not illustrated)

(Black-bellied Terrapin)

*Pelusios seychellensis* Torti Soupap

Length: up to 26 cm/10 in

Probably extinct, known only from three specimens collected on Mahé in 1895. Some Seychelles Yellow-bellied Mud Turtles may show features of this species and potential hybridisation has raised doubts over the continued validity of the species. All terrapins in Seychelles are endangered and remain poorly known.

CONSERVATION STATUS: **Extinct**

**IDENTIFICATION:** Similar to Seychelles Yellow-bellied Mud Turtle but differs in its entirely black, not yellow, underside (plastron).
**DESCRIPTION:** A small turtle with a domed blackish-olive carapace, black below. The skin is olive-grey, yellower on the chin and throat, and with wavy brown and yellow lines on top of the head.

1

1

2

2

GECKOS

## E Stripeless Day Gecko 1

*Phelsuma astriata* Lezar Ver

Length: 60 mm/2·4 in (excluding the tail)

A widespread small day gecko, found on trees and shrubs but also occurring in and around houses. On Praslin it is less common in the lowlands and more frequent in the hills.

**IDENTIFICATION:** The day geckos are all very similar, being largely bright green above with various reddish markings, and whitish below. Range is a useful criterion when attempting to identify individuals. The Stripeless Day Gecko is smaller than Sundberg's Day Gecko and lacks the red chevron on the chin of both that species and the similar-sized Long Island Day Gecko.
**DESCRIPTION:** Bright green above with irregular patches of dark red on the back and sometimes smaller patches on the tail; there is a red chevron between the eyes and usually two bars across the neck. Individuals on some islands show a distinct red line down the centre of the back. The sides are uniformly green or greyish-brown and lack bold markings. The underside is uniformly creamy-white. The eyes are black. **BEHAVIOUR:** All day geckos are well camouflaged for their daytime existence on trunks and stems of plants. Stripeless Day Geckos often favour palms and exotic shrubbery's in hotel gardens, but will also occur inside houses where they sometimes feed around lights in the evening. The nectar of Banana flowers attracts day geckos of all species.
**BREEDING:** Breeds throughout the year, females laying two small (10 mm in diameter) round, white eggs under bark or in a hole in a tree, which hatch after around 70 days.

## E Long Island Day Gecko 2

*Phelsuma (sundbergi) longinsulae* Lezar Ver

Length: max. 67 mm/2·6 in (excluding the tail)

A widespread small gecko of variable size. Individuals on Mahé, Bird and Thérèse are larger, stockier and broader-headed than those on Frégate, Silhouette and other small islands around Mahé. It does not occur on the same islands as Sundberg's Day Gecko; these two forms may be races of a single species.
It occurs in similar habitats to the Stripeless Day Gecko.

**IDENTIFICATION:** Similar in size to the Stripeless Day Gecko from which it is best separated by the presence of one and sometimes two dark chevrons across the neck, a red chevron on the back of the head and a more pronounced red line from the nostril to the eye. It also has a dirty white underside, mottled with purple, and often a yellowish vent.
**DESCRIPTION:** Bright green above with one to three rows of spots running down the back which may join together; there are usually red bars across the tail. The patterning around the head and the colour of the underparts is described above. The eyes are black. **BEHAVIOUR:** Much as Stripeless Day Gecko. **BREEDING:** Much as Stripeless Day Gecko.

## E Sundberg's Day Gecko 3

*Phelsuma sundbergi* Lezar Ver

Length: max. 100 mm/3·9 in (excluding the tail)

A widespread relatively large green gecko that does not occur on the same islands as Long Island Day Gecko; it is possible that the two forms may be races of a single species.
It occurs in similar habitats to other day geckos, but on Praslin is more common in the lowlands and less frequent in the hills.

**IDENTIFICATION:** Adults are normally obviously larger, stockier and broader-headed than Stripeless Day Gecko. In addition, they have one, and sometimes two, dark chevrons across the neck, a faint red chevron on the back of the head, a red line from the nostril to the eye, and a dirty white underside, often with a yellowish wash.
**DESCRIPTION:** Bright green above, often with a yellowish wash, with an even scattering of red spots lacking any obvious longitudinal pattern. The tail is spotted red and is never barred. The patterning around the head and the colour of the underparts is described above. The eyes are black.
**BEHAVIOUR:** Much as other day geckos, but especially fond of garden shrubberies.
**BREEDING:** Much as other day geckos although the paired round, white eggs are larger.

1

2
2

3

## 1 Pacific Gecko 1

*Gehyra mutilata* Lezar Disik

Length: 55 mm/2·2 in (excluding the tail)

The most frequently observed gecko in Seychelles, often occurring inside houses and hotels, where they cluster around lights at night. A native of the Pacific region, this commensal of man has spread right round the world. It is very common on the larger islands and has recently colonised smaller offshore islands such as Cousine.

**IDENTIFICATION:** A relatively small gecko of very variable colour, best identified by its generally pale pinkish-brown or grey, almost-transparent, appearance lacking obvious markings, and its large, rounded toe pads and relatively large, flattened tail which is constricted at the base. The bold blackish eyes are distinctive. The similar Common House Gecko is restricted to Bird Island.

**DESCRIPTION:** The ground colour of the body is highly variable and can be changed rapidly to suit the background or reflect the mood of the individual. The upperparts may be dark brown when foraging on trees, but are normally light grey, brown or pinkish when foraging close to lights in buildings. They are covered with small black and white spots above but these all but disappear in strong light. The underparts are yellowish or white. The eyes are conspicuously black, and induce a 'goggle-eyed' appearance. The toe pads are large and pale, whilst the flattened tail is relatively large and is obviously constricted at the base. Immatures are generally darker brown with some bold white spots **(1i)**.

**VOICE:** A noisy species, which frequently calls a repeated series of similar "*chek*" notes.

**BEHAVIOUR:** Strictly nocturnal in habits, appearing at dusk from behind pictures, cracks in walls or natural crevices in trees, to feed on insects. Often found clustered high on walls or on ceilings to prey on insects attracted to lights.

**BREEDING:** Females lay two round, white eggs in cracks, which hatch after about 60 days. Breeding occurs throughout the year.

## 1 Common House Gecko 2

*Hemidactylus frenatus* Lezar Disik

Length: 55 mm/2·2 in (excluding the tail)

A widespread commensal species, currently restricted in Seychelles to Bird Island where it occurs alongside, but greatly outnumbers, the very similar Pacific Gecko.

**IDENTIFICATION:** A relatively small gecko of variable colour, best separated from the similar Pacific Gecko by its broader head (often held up), proportionately shorter, slimmer tail lacking a constricted base, the presence of a dark streak from the nostril to the eye, dark mottling above on some individuals, vertical pupil slits, smaller pads on slimmer toes and a much reduced first toe on each foot.

**DESCRIPTION:** The ground colour of the body is highly variable, which, as with Pacific Gecko, can be changed rapidly to suit the background and mood of the individual. The upperparts may be grey or pinkish-brown above with scattered darker flecks and bars on the tail, or uniformly pale yellowish or pinkish-white. There is sometimes a dark orange-buff line from the nostril to the eye and an orange-buff stripe along the flank. The underparts are white. The eyes are bronzy-orange with a vertical pupil slit that widens at night. The toe pads are relatively small, elongated and pale, with a prominent central claw. The flattened tail is relatively slim and short, and has a spiky ridge of scales on either side near the base.

**VOICE:** A very vocal species, emitting a distinctive chirping call "*chit chit chit-chat-chat*", mostly during the night and less often during the day.

**BEHAVIOUR:** Largely nocturnal in habits, hiding during the day in crevices in buildings, under bark or among stones, and emerging at dusk to feed on insects. On Bird Island, it occurs in large numbers in the hotel chalets and feeds actively around wall lights. This species is a more active hunter than the Pacific Gecko, often sitting with its head raised and then darting off rapidly to catch prey.

**BREEDING:** Females lay two round, white eggs in cracks, which hatch after about 65 days.

## E Seychelles Sucker-tailed Gecko (3)

*Urocotyledon inexpectata* Lezar Disik

Length: 80 mm/3·1 in (including the tail)

A widespread but poorly-known species which is rarely seen owing to its small size and cryptic habits. It inhabits crevices in rocks, trees and buildings, rarely venturing far in the open.

**IDENTIFICATION:** A small gecko with long toes, each distinctly tipped on the underside with a pair of pads used for adhesion, and with white adhesive organs on the underside of the tip of the tail. Much more cryptic in habit than the Common House Gecko, and more boldly marked above. Always appears slimmer, with a proportionately larger head set on a thinner neck, than immature Bronze Gecko. **DESCRIPTION:** Very variable in colour. Males are normally brown or grey above with bold alternating bands of black and white from the neck to the tail tip. Some females at least are a paler sandy-brown and less boldly marked with pairs of darker spots in rows from the nape to the base of the tail; they are whiter below, often yellower on the mid-belly. **BEHAVIOUR:** Lurks in crevices or under bark, where it is well-camouflaged and hard to spot. Probably more active at night, but also observed during day. **BREEDING:** Poorly known; though an egg-bound female was observed on Aride in August. Lays a pair of small, white eggs in rock or tree crevice.

## E Bronze Gecko ①

*Ailuronyx seychellensis* Lezar Mangouya

Length: 200 mm/7·9 in (including the tail)

A large gecko, restricted to areas of more natural forest, particularly palm forest on the larger islands, but more widespread on smaller rat-free offshore islands where they also occur inside buildings.

**IDENTIFICATION:** A large, stocky arboreal gecko, smaller than the very local and poorly known Giant Bronze Gecko *(page 110)*, but very similar in appearance to the slightly smaller Dwarf Bronze Gecko, from which best separated by its shorter, blunter and more steeply-angled snout. The eyes are also relatively small, are more sunken in their sockets and appear to be placed further forward on the snout compared with Dwarf Bronze Gecko.

**DESCRIPTION:** A large gecko, which exhibits an extensive range of coloration. Individuals inhabiting buildings are often smoother in appearance, paler in colour and lack bold markings, whilst those living amongst rocks or trees tend to show a rougher skin, and are darker and more strongly marked. Individuals can change colour rapidly, and project their body scales to give a rough-skinned look. The body colour is variably brown, bronze, olive-grey, or even blackish, and may be largely devoid of stronger markings in some individuals, or boldly marked with dark and white spots and bars in others. There is often a pronounced pale band with dark borders leading from the snout via the eye towards the shoulder, and the tail is often barred with alternating dark and pale bands. Some individuals show a continuous dark-bordered creamy band along the sides from the eyes and joining across the base of the tail. The underside is a dirty or creamy white. The eye is large and bronze-coloured with a vertical pupil slit. The toes all bear elongated pads.

**VOICE:** Generally silent, but individuals may squabble loudly emitting harsh gurgling noises.

**BEHAVIOUR:** Although largely nocturnal, some individuals are active throughout the day. During the day they rest on the trunks of trees, the fronds of palms and in rock crevices. At night, they feed on insects on tree trunks, and on the nectar from banana flowers, but some also feed on the ground in rocky areas on rat-free islands and in buildings where they feed on insects attracted to lights.

**BREEDING:** Females lay two large round white eggs in depressions on tree-trunks, in rock clefts or on walls. These hatch after about 70 days and the female often remains close by before they hatch.

## E Dwarf Bronze Gecko ②

*Ailuronyx tachyscopaeus*

Length: 84 mm/3·3 in (excluding the tail)

A recently discovered species that is common in lowland coconut scrub on Silhouette. Elsewhere it is known from a few sites on Mahé and Praslin, where it seems to prefer remnant patches of native palm forest, such as La Reserve (Mahé) and Vallée de Mai (Praslin) (where it appears to outnumber Bronze Gecko). The Dwarf Bronze Gecko has also been found inside buildings on Mahé and on La Digue.

**IDENTIFICATION:** Very similar in both structure and coloration to Bronze Gecko. Best separated by its smaller adult size and longer, more tapering snout which gives the species a thin-headed appearance when viewed in profile. The eyes are also proportionately larger and bulge more from their sockets compared with the Bronze Gecko.

**DESCRIPTION:** A mid-sized gecko exhibiting the same extensive range of coloration as the larger Bronze Gecko. The body colour is variably brown, bronze or olive-grey, and may be largely devoid of stronger markings in some individuals, or boldly marked with dark and white spots and bars in others. There is often a pronounced pale band with dark borders leading from the snout via the eye towards the shoulder, and the tail is often barred with alternating dark and pale bands. The eye is large and bronze-coloured with a vertical pupil slit.

**BEHAVIOUR:** Predominantly nocturnal in habit, resting on lower trunks of palms or hidden in foliage during the day.

**BREEDING:** Poorly known.

1
2
1
1

## Ⓔ Giant Bronze Gecko ①

*Ailuronyx trachygaster*

Length: 260 mm/10·2 in (excluding the tail)

An extremely large and poorly-known gecko. Although originally named from a single 19th century specimen, supposedly from Madagascar, geckos closely matching the description of this species are regularly seen in mature Coco-de-Mer forest in the Vallée de Mai and Fond Ferdinand on Praslin, and more rarely in native plam forest on Silhouette.

**IDENTIFICATION:** A very distinctive gecko; adults being easily told from the bronze geckos by their immense size, their broad, angular, wedge-shaped head, very large feet with broad white toe-pads and thick, powerful hind-legs.

**DESCRIPTION:** A massive gecko, uniformly rich amber or olive-brown above. The body is thickset with particularly sturdy hind-legs and there are large flaps of loose skin behind the head and the back-legs. The broad, angular head is characteristic, as is the distinctly 'roman' nose and depression between the nostrils. The feet are large and each toe has a large rounded pad, which is white on the underside. The tail is thick-based and relatively long. The eyes are large.

**BEHAVIOUR:** Hard to spot in the dark shade of the forest, adults can be found lurking during the day high up on mature Coco-de-Mer nuts where they blend perfectly with the outer sheaths of the nut. Often only the dull glow of their eyes is visible to the unaided eye. They creep slowly over the surface of the nuts, laboriously raising one foot in turn and re-attaching the large toe-pads, sometimes stopping to sip water running over the nut.

**BREEDING:** Completely unknown.

CHAMELEON

## Ⓔ Tiger Chameleon ②

*Chamaeleo tigris* Kanmeleon

Length: 160 mm/6·3 in (including the tail)

The only chameleon in Seychelles and restricted to the three largest islands, where it occupies wooded habitats from sea level to the higher peaks. Generally scarce or, at best, elusive, it is well camouflaged for its life in the trees and is only rarely encountered on the ground.

**IDENTIFICATION:** A typical chameleon, with a wobbly, slow-moving gait, rotating 'turret-like' eyes and the ability to change colour rapidly. It is unlikely to be confused with any other species.

**DESCRIPTION:** The sexes of the Tiger Chameleon are similar although females are normally larger than males. It is a relatively small and slim species which, although able to change colour rapidly to suit its background, is typically yellowish-brown with darker transverse stripes and blotches. It can also appear completely black when disturbed. The tail is quite long and curls around vegetation to provide additional support. Each limb bears two claw-like fingers, which grasp branches and stems. There are rows of short spines on the head, throat and back.

**BEHAVIOUR:** Moves very slowly through vegetation with deliberate movements of the limbs. It characteristically shakes and wobbles its body as it moves, rotating its turret-like eyes independently of each other.

**BREEDING:** Poorly known, although it is believed that breeding occurs at the start of the northwest monsoon (October–November), the females laying up to a dozen oblong, white eggs in moist soil.

2
1

SKINKS

## E Seychelles Skink (1)

*Mabuya sechellensis* Lezar Sek

Length: 215 mm/8·5 in (including the tail)

The most widespread and numerous terrestrial lizard in Seychelles. Occurs in most habitats up to 500 m (1,600 ft), but is more common in lowland areas, and particularly on seabird islands such as Aride and Cousin where they reach very high densities.

**IDENTIFICATION:** The only common terrestrial diurnal lizard found on most islands. This species is slimmer and more boldly marked than the larger Wright's Skink, with a darker brown back and a bold blackish line along the sides from the eye to the back leg that is bordered above and below by a thinner pale whitish stripe. The other skinks are smaller, more uniformly coloured and cryptic in habits, and lack the long slim tail.

**DESCRIPTION:** A neatly-proportioned, slim, long-tailed skink which is mid- to dark brown above, flecked with darker and sometimes paler spots. There is always a bold black stripe along the side from the eye to the hind-legs, bordered above and below by paler whitish lines. The underside is variably whitish to bright yellow, usually with a whiter throat. The beady dark eye has a thin golden eye-ring.

**BEHAVIOUR:** Diurnal in habits, favouring open sunny areas, where it runs rapidly across the ground, but also climbs rocks and trees. It can often be heard rustling through dry leaf-litter before being seen. Although relatively shy on the more heavily populated islands, Seychelles Skinks can be very tame and inquisitive on smaller offshore islands, often coming to humans in search of food. They feed on a variety of insects and fruit as well as dropped fish, carrion and other smaller skinks.

**BREEDING:** Breeds throughout the year, the females burying two to six white, leathery, oblong eggs in the soil or under stones; these hatch after around 60 days.

## E Wright's Skink (2)

*Mabuya wrightii* Lezar Tengteng

Length: 290 mm/11·4 in (including the tail)

A distinctive large terrestrial lizard, restricted to rat-free seabird islands, where it can reach impressive densities.

**IDENTIFICATION:** A large, terrestrial diurnal lizard, more thickly-set and less boldly marked than the smaller and more widespread Seychelles Skink. It is variably grey-brown above, mottled with paler spots, and never shows a pronounced blackish lateral stripe bordered with paler lines as in Seychelles Skink (although some younger individuals may show a diffuse dark line along their sides). The black stripe along the back of the back-legs is distinctive.

**DESCRIPTION:** A large, stocky lizard with a proportionately shorter tail than Seychelles Skink. They are greyish-brown above, often with a shiny pale greenish tint, and liberally spotted black and pale blue-grey. There is sometimes a diffuse dark stripe along the side but this is never as pronounced as in Seychelles Skink and is not bordered by obvious pale lines. The underside is normally a very pale creamy yellow, always with a whiter throat. The large, dark eyes are encircled by a thin golden eye-ring.

**BEHAVIOUR:** Diurnal in habits, and rather more terrestrial than the Seychelles Skink (although it does climb rocks and trees). This species frequents the shaded forest floor on seabird islands but also occurs in grassland on Recif, basking in the sun on favoured rocks and trunks during the morning. It is generally quite tame and inquisitive and will come to humans in search of food. It feeds on a variety of insects and fruit as well as dropped fish, carrion, smaller skinks and even small birds.

**BREEDING:** Wright's Skink breeds throughout the year, the females burying 2–6 white, leathery, oblong eggs in soil or under stones.

1
1
2
2

## E Brauer's Skink

1

*Janetaescincus braueri* Lezar Fen
(includes former Vesey Fitzgerald's Skink *Janetaescincus veseyfitzgeraldi*)

Length: 53 mm/2·1 in (excluding the tail)

A small, thin skink with tiny legs and a pointed snout. Widespread, inhabiting leaf-litter of high elevation forest on Mahé but also more commonly in woods at lower levels on Silhouette, Frégate, La Digue and Félicité.

**IDENTIFICATION:** A small skink, smaller than Gardiner's Skink, with a much slimmer body and a more pointed snout, and considerably smaller than Seychelles Skink *(page 112)*, with proportionately shorter legs and tail. The front feet bear only four toes, instead of five, a feature that separates this species from all other skinks in Seychelles.
**DESCRIPTION:** A small, slender diurnal lizard with tiny legs and a pointed snout. The upperparts are a rather uniform dark brown, sometimes bearing paler markings, and can appear iridescent in good light. The underparts are light yellow, often with a smudging of brown.
**BEHAVIOUR:** Diurnal in habits and usually observed scuttling rapidly through the leaf-litter of forests at lower elevations. It also lives deep in the leaf-litter and amongst the tangled roots of montane forest on Mahé, where it is rarely observed.
**BREEDING:** Females lay two soft, white, leathery, oblong eggs in moist cavities in soil or under piles of coconuts.

## E Gardiner's Skink

2

*Pamelaescincus gardineri* Lezar Gra

Length: 67 mm/2·6 in (excluding the tail)

A squat, flattened skink with short, stumpy limbs, which generally lives beneath logs and in leaf-litter and is rarely seen on the surface. It is a widespread species which, on Mahé, occurs from sea level to at least 600 m (2,000 ft).

**IDENTIFICATION:** A squat, terrestrial lizard with shortened limbs (particularly the fore-legs) and a relatively short thick-based tail. Gardiner's Skink rarely appears on the surface and is therefore unlikely to be confused with other species of skink. It appears very shiny and smooth above and when exposed moves with a rapid squirming action, not unlike a fish, rapidly disappearing from view into the leaf-litter. Seychelles Skink *(page 112)* is much longer-tailed, slimmer and more boldly marked. Wright's Skink *(page 112)* is larger, with stronger limbs and a slower gait. Brauer's Skink is much slimmer with a fine-pointed snout.
**DESCRIPTION:** A mid-sized, flattened lizard with reduced limbs and a proportionately short tail. It is variably light to dark brown above with darker flecks. The sides bear a diffuse dark stripe which hs some pale spotting and runs from the head towards the back legs. The underside is yellow and tinged pink on the tail (although young animals are pinker below). The eyes are dark and relatively large.
**BEHAVIOUR:** The secretive habits of this species can be a distinctive feature, the animals usually hiding under logs or in leaf-litter during the day. It is more active at night and during rain when it comes to the surface to feed on insects, although may be more active during the day on the larger islands. The rapid 'swimming' motion described above enables rapid escape from disturbance.
**BREEDING:** Breeds throughout the year, the females burying two white, oblong eggs in the soil or under stones in communal nests.

1

2

SNAKES

## Ⓔ Seychelles House Snake ①

*Lamprophis geometricus* Koulev Gri

Length: up to 90 cm/35 in

A thick-bodied nocturnal snake that typically hunts on the ground but also climbs trees. Widespread and fairly common in wooded habitats as well as in more open rocky areas and in villages from sea level to mid elevations but apparently absent from montane forest.

**IDENTIFICATION:** Readily separated from the Seychelles Wolf Snake by the much thicker and shorter body, larger head with broad, truncated snout, small bronze-gold eyes with characteristic black vertical slit pupils, pronounced white stripes behind the eyes, and a smooth appearance to the body created by the absence of keeled scales along the back. **DESCRIPTION:** A thick-bodied relatively short snake, which is typically brown above with three dark stripes down its back; in good light its smooth skin shines with multi-coloured tints. The underside is grey with paler narrow whitish bands. The head is brown with a whitish snout and a distinctive white stripe running from the back of each eye towards the neck. **BEHAVIOUR:** Nocturnal in habits, killing its prey of mice, rats and lizards by constriction. It has been recorded entering houses at night and will strike out aggressively if cornered. **BREEDING:** Poorly known.

## Ⓔ Seychelles Wolf Snake ②

*Lycognathophis seychellensis* Koulev Zonn

Length: up to 130 cm/51 in

A very slim diurnal snake that typically hunts on the ground, although it can also climb trees. Widespread and fairly common in most wooded habitats from sea level to mountain-top, but relatively shy and unobtrusive.

**IDENTIFICATION:** Much slimmer than the Seychelles House Snake, the small head having a relatively pointed snout and a broad rear-edge. The large golden-orange eyes have rounded pupils and there are no pronounced white stripes behind the eyes. The skin appears rough due to the presence of keeled scales along the back. **DESCRIPTION:** A slim snake, which comes in three distinct colour phases: blackish **(2a)**, olive-yellow with brown smudges **(2b)** and an intermediate phase of grey-brown with darker markings which become progressively blacker towards the tail **(2c)**. Olive-yellow snakes are bright yellow on the underside whilst blackish and intermediate-phase snakes are normally whitish below with frequent darker smudges. There is often a dark stripe along the side of the head passing through the eye, giving a masked appearance. **BEHAVIOUR:** Most active early in the morning and again towards dusk, stealthily moving across woodland floor searching for small skinks, geckos, frogs and small birds. Often lies motionless alongside fallen logs, but also climbs trees in search of geckos. Relatively shy, slinking away from human approach, often disappearing into holes or hollow logs. Less aggressive than Seychelles House Snake and often very calm when handled. **BREEDING:** Breeds throughout the year, the females laying two leathery, oblong, white eggs under logs or coconut husks.

3

## 1 Brahminy Blind Snake 3

*Ramphotyphlops braminus* Lev-d-ter Nwar

Length: up to 17 cm/6·7 in

A very small worm-like snake that burrows through moist soil and rotting logs, and is often accidentally transplanted in the soil of potted plants. An introduced species that is now widespread in lowland habitats on most islands. The cryptic habits and small size of this species make it very unobtrusive.

**IDENTIFICATION:** The tiny rounded body is more worm-like than snake-like, and its dark coloration could lead to confusion with millipedes. It is separated from both by the presence of a small forked tongue, which darts in and out when an animal is captured, the finely scaled body and the presence of tiny vestigial eyes. **DESCRIPTION:** A very small snake with a rounded body, blunt head and more pointed tail. The upperparts are normally blackish above and the snout and underside are paler and greyer. **BEHAVIOUR:** Burrows through soil in search of invertebrate prey and is rarely seen on the surface except after heavy rain. Most often discovered in soil of potted plants. Regularly scents its surroundings by flicking its tongue in and out. **BREEDING:** All individuals of this species are female, which are capable of laying eggs without being fertilised by a male. Each female lays clutches of up to seven eggs in moist soil which hatch out into exact replicas of their mother.

# AMPHIBIANS

Amphibians are well represented in Seychelles, with six species of frogs and seven species of primitive burrowing caecilians. The degree of endemism is high, with only one of the 13 species not being endemic. This is the Mascarene Frog, which also occurs on other islands in the Indian Ocean and is a widespread and common species that has adapted well to man-made changes in habitats. The caecilians remain poorly known because of their cryptic life-style but appear to be relatively widespread in the lowlands of their respective islands. Sadly, the endemic frog species have fared less well. The four remarkable small endemic frog species, with their widely different breeding strategies, are restricted to tall moist forest on the larger islands. The rarest, Thomasset's Frog, is restricted to higher, wetter patches of mostly native vegetation on Mahé and Silhouette. The Seychelles Tree Frog is also most common in the hills although it does occur in the lowlands.

## E Seychelles Tree Frog (1)

*Tachycnemis seychellensis* — Krapo

Length: male 45 mm/1·8 in; female 70 mm/2·8 in

Generally scarce in the lowlands of the four largest islands but not uncommon in vegetation close to streams at higher elevations.

**IDENTIFICATION:** The largest frog in Seychelles and the only one which regularly rests on leaves and branches in shrubs and trees, using sucker-like pads on the tips of their toes. The sexes are distinct, females being larger and greener, and males smaller and yellow-brown. The tadpoles are quite large, up to 65 mm in length.

**DESCRIPTION:** Females **(1f)** are obviously larger than males **(1m)** and are generally light to dark green above, whilst males are yellowish-brown to dark-brown above. Both sexes have large, whitish eyes. The underparts of males and smaller females are a pale creamy yellow, but larger females have a dark reddish underside, often separated from the green back by a band of yellow. On Praslin, both sexes often have pale spots on their backs. Rounded pads on the end of each toe are used for climbing.

**VOICE:** Males produce a characteristic series of duck-like "*quack*" notes and also a softer "*tuc*". They sing in chorus year-round in hill valleys, but generally only during wet weather in the lowlands.

**BEHAVIOUR:** Mostly active at night and during wet weather, adults often rest on the underside of large leaves in moist, shady spots during the day **(1r)**. Moves by crawling across vegetation rather than hopping. Occasionally individuals can be found in domestic bathrooms.

**BREEDING:** Occurs throughout the year, but most frequently during the wetter northwest monsoon (December–February). Females lay small piebald eggs on low vegetation above or close to a stream, and the tadpoles fall or crawl into the water where they live for several months before transforming into froglets.

## Mascarene Frog (2)

*Ptychadena mascareniensis* — Grenuiy

Length: 45–60 mm/1·8– 2·4 in

A widespread species that is common in most moist habitats from lowland wetlands and drainage ditches to montane forest.

**IDENTIFICATION:** The only typical ground-dwelling frog in Seychelles, appearing large-headed with large eyes, smooth skin and long back legs. Does not climb into vegetation like the Seychelles Tree Frog, and has webbed, pointed toes which lack the rounded pads of that species. Tadpoles are small $<$40 mm in length.

**DESCRIPTION:** Mid olive-brown above, with dark spots forming a bold wedge from the pointed snout down the back. There is darker striping on the lower back and legs, and often a narrow white line along the upper surface of the hind legs. The underside is yellowish-cream and unmarked.

**VOICE:** Males sing in loud chorus at night and during heavy rain, emitting a hoarse "*wah wah wah*".

**BEHAVIOUR:** Typically found in wet grassy areas close to pools and streams, where it is active throughout the day. Animals are capable of leaping large distances using their long back legs.

**BREEDING:** Occurs throughout the year, but most frequently during the wetter northwest monsoon (December–March). Females lay their small piebald eggs singly, directly into standing water, including roadside puddles. The eggs hatch within a few days to release small tadpoles, which metamorphose into froglets in less than a month.

1m
1r
1f
1

2

## E Thomasset's Frog

1

Length: 45–55 mm/1·8–2·2 in

(Thomasset's Tocking Rock-frog)

*Nesomantis thomasseti* Grenuiy

The rarest frog species in Seychelles. It is rarely seen and poorly known, inhabiting rocky valleys in high forest areas (usually above 350 m (1,000 ft)) on Silhouette and Mahé.

**IDENTIFICATION:** The largest of the four smaller endemic frogs, as large as Mascarene Frog, but golden to reddish-brown above with scattered black markings. **DESCRIPTION:** The sexes are similar although females are larger than males. Golden to reddish-brown above with darker markings, often with a pale line down the middle of the back extending onto the thighs, and a white line extending backwards from each eye towards the front legs. There are black bars across the tops of the legs and toes, and a double row of light coloured bumps down the back starting behind each eye. The underside is brown with paler mottling. The large eyes are golden with black pupils. **VOICE:** Rarely heard, males calling singly during wet or low-cloud conditions day or night. The call is similar to Seychelles Frog, but is longer and slower, with more than one guttural *"raack"* note preceding the series of *"toc"* notes e.g. *"raack–raack–raack–toc–toc–toc–toc"*. **BEHAVIOUR:** Climbs well over rocks and usually observed sitting on boulders or occasionally on low branches at night. **BREEDING:** Poorly known but thought to lay eggs on land that hatch directly into froglets, since tadpoles have not been found.

## E Seychelles Frog

2

Length: 15–20 mm/0·6–0·8 in

(Croaking Carry-cot Frog)

*Sooglossus sechellensis* Pti Grenuiy

More restricted in range than Gardiner's Frog, preferring areas of wetter, less disturbed forest at higher elevations on Mahé and Silhouette, although occurs as low as 200 m (650 ft). It only calls infrequently and is therefore hard to locate, but is not uncommon in suitable habitat.

**IDENTIFICATION:** Intermediate in size between Gardiner's and Thomasset's Frogs. Best separated from the more widespread Gardiner's Frog by its larger size, lack of a dark band on the side of the body and head, and very different call. Could be confused with young of the larger Thomasset's Frog but lacks the thin white stripe down the centre of the back which is typical of the young of that species. **DESCRIPTION:** Females are larger than males but are similar in colour. The back, sides and fore legs are golden-brown with black spots and bands. There is a large triangular-shaped black spot on the top of the head behind the eyes (though note that Thomasset's Frog may also show this feature). **VOICE:** Single males call only rarely from hiding places in the leaf-litter. Sporadic calls may be heard day and night but most often during wet weather. The call is similar to Thomasset's Frog but usually consists of only one guttural *"craack"* note, which is generally (but not always) followed by three or four *"toc"* notes. **BEHAVIOUR:** Hides in leaf-litter on the forest floor and is not easy to see. **BREEDING:** Breeds mostly during the wetter northwest monsoon. Females lay eggs in nests under leaves and sit on them for up to three weeks until they hatch as small tadpoles. The tadpoles then climb onto their mother's back where they transform into small froglets without eating; they then hop off to pursue independent lives.

## E Palm Frog

3

Length: 12–15 mm/0·5–0·6 in

*Sooglossus pipilodryas* Pti Grenuiy

A recently discovered species, apparently restricted to areas of native palm forest on Silhouette, where not uncommon. Has been searched for, without success, in similar habitat on Mahé.

**IDENTIFICATION:** Slightly larger than the more widespread Gardiner's Frog from which best separated by its longer legs, broader body and distinctive dark diamond shape on a brown back. It also lacks the dark lateral band on the side of the body typical of that species.
**DESCRIPTION:** Females are larger than males, but are similar in colour. The back, sides and upper legs are typically brown bearing a distinctive black diamond shape on the back, although some individuals are a reddish-brown and others are striped darker above. **VOICE:** Call is a high-pitched *"ppep"* usually repeated four times in a row, and thus similar to call of Gardiner's Frog. **BEHAVIOUR:** Typically found hiding in the leaf axils of endemic palms, but also occurs in adjacent leaf litter. **BREEDING:** Unknown.

## E Gardiner's Frog 4
### (Pygmy Piping Frog)
### *Sooglossus gardineri* — Pti Grenuiy

Length: 11–13 mm/0·4–0·5 in

The smallest frog species in Seychelles, if not the world. It is abundant in the damp rocky understorey of mountain forest above 200 m (650 ft) on Mahé and Silhouette, where it is frequently heard but very hard to see without diligent searching.

**IDENTIFICATION:** The smallest of the three endemic sooglossid frogs. It is highly variable in coloration but readily identified by its tiny size and the dark sides to the head and body which always contrast with the paler upper and lower surfaces. **DESCRIPTION:** Females are larger than males, and both are highly variable in colour, ranging from light tan to reddish-brown above. Some have a white stripe down the back from the head to the hind legs, and the back may bear dark or pale blotches. However, all have dark sides to the head and body forming a lateral band which contrasts with the back and the moderately dark underside. **VOICE:** Single males emit high-pitched *"peep"* notes, particularly in damp weather during the day, when they appear to be everywhere! Calls less often at night. **BEHAVIOUR:** Hides under leaves, in moss and in hollow stems. The somewhat ventriloquial calls stop when the frog is approached too closely, making the species very hard to see. **BREEDING:** Breeds throughout the year, but particularly during the wetter northwest monsoon. Females lay eggs in nests under leaves, and sit on them for up to a month, guarding them from predators and keeping them moist. There is no aquatic tadpole stage and the eggs hatch directly into froglets.

# CAECILIANS

Lev-d-ter Nwar

Caecilians are curious burrowing amphibians with elongated, legless bodies, giving them a worm-like appearance. They live in damp soil and leaf-litter, and are rarely seen unless actively searched for under logs or in piles of rotting vegetation. One species, the Swimming Caecilian, also enters streams and may be found swimming eel-like or burrowing into the stream bed.

**DESCRIPTION:** Caecilians have small, pointed snouts, with the mouth usually situated well back from the tip. The eyes are small and much reduced. Instead, there is a pair of sensory tentacles on either side of the snout, which are used for exploring their environment. The body is divided into many rings, or annuli, which may be further sub-divided into secondary or even tertiary rings.

**BREEDING:** Large white eggs are laid in strings into damp soil cavities. The female lies with the eggs protecting them until they hatch. All but one species produce small larvae, which wriggle towards streams where they live in an aquatic form until they eventually metamorphose into terrestrial caecilians. Eggs of the Swimming Caecilian hatch directly into small terrestrial caecilians.

**IDENTIFICATION:** Seven species of Caecilian occur in Seychelles. All are very similar in appearance and require close views to allow accurate identification. The key features are summarised below:

### E 1 Big-headed Caecilian *Grandisonia alternans*

Length: 280 mm/11 in. A large caecilian with a distinctively large head, rather blunt snout, and prominent eyes; the tentacles are positioned roughly halfway between the eyes and the snout. All annuli are divided into secondary rings.

### E 2 Flat-headed Caecilian *Grandisonia larvata*

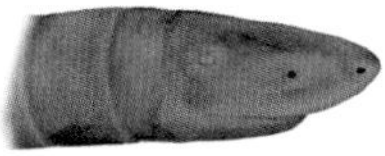

Length: 205 mm/11 in. A mid-sized caecilian with a rather flat top to the head and tentacles positioned roughly half way between the eyes and the nostrils. The first few annuli behind the head are not subdivided into secondary rings.

### E Praslin Flat-headed Caecilian *Grandisonia diminutiva*

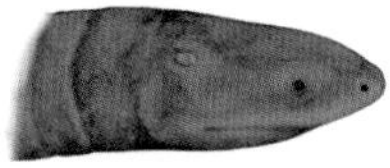

Length: 205 mm/11 in. Occurs only on Praslin where it replaces the very similar, and possibly conspecific, Flat-headed Caecilian.

### E Pin-headed Caecilian *Grandisonia sechellensis*

Length: 220 mm/9 in. A mid-sized caecilian with a small head, rather rounded snout and tentacles positioned closer to the eye than to the tip of the snout. The first few annuli behind the head are not subdivided into secondary rings.

### E 3 Small-headed Caecilian *Grandisonia brevis*

Length: 120 mm/5 in. The smallest caecilian with a distinctive small head and very pointed snout, very small eyes and tentacles positioned towards the tip of the snout, very close to the nostrils. All annuli are divided into secondary rings. Only found on Mahé and Silhouette.

### E Swimming Caecilian *Hypogeophis rostratus*

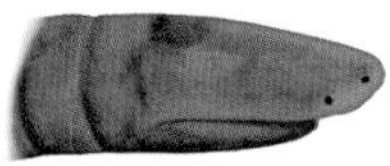

Length: 365 mm/14 in. The largest and most commonly seen caecilian. It has a relatively small head with the tentacles positioned far forward on the snout. Most of the annuli are not sub-divided into secondary rings, other than close to the tail-tip.

### E Toothy Caecilian *Praslina cooperi*

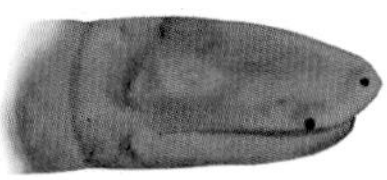

Length: 230 mm/19 in. The rarest caecilian. It has a distinctive mouth at the end of the snout containing more teeth than the other caecilians. The tentacles are positioned just in front of the eyes. The first few annuli behind the head are not subdivided into secondary rings.

1
2
3
1
2
3

# INVERTEBRATES

Seychelles is home to a very wide variety of invertebrates and it is well beyond the scope of this book to provide the means of identifying them all. Detailed sections are included on butterflies plus the two commoner hawk-moths and on the damselflies and dragonflies. The final section aims to help those with an interest in other insects, millipedes and centipedes, whip scorpions, spiders, crabs, and slugs and snails to identify the more conspicuous and frequently encountered species.

Many of the endemic species which were first recorded in the nineteenth century, when habitat change was less marked, have either been lost or are now conspicuously rare. On the other hand, new species continue to be found and the current list of species known from Seychelles is certainly far from complete.

## Insects

### Butterflies & moths (Lepidoptera)

*(pages 126–132)*

The lepidopteran fauna of Seychelles is relatively poorly known, and the average observer is struck by the conspicuous absence of large showy butterflies so typical of many tropical countries. We can only guess as to whether this is a result of wide-scale habitat alteration and loss. Only eleven species of butterfly and two species of hawk-moth are at all likely to be seen, although many other migrant species, together with presumed scarce resident species, have been recorded from time to time. The group is dominated by cosmopolitan species such as the African Grass Blue, and by strongly migratory species whose resident populations on the islands are probably augmented regularly by new arrivals. There is only one endemic species of butterfly, the King of Seychelles, which is rarely seen and probably endangered in the high forests on Mahé and Silhouette.

Easily the most numerous and widespread species of butterfly in Seychelles is the African Grass Blue, which darts rapidly low over low vegetation such as grass lawns and rarely settles to allow a close look. In shrubbier areas of leguminous plants, they are often joined by the larger Long-tailed Blue, although the very similar Lang's Short-tailed Blue is also present. The Tiny Grass Blue also occurs locally in areas of tall grass. One species of skipper, the Olive-haired Swift, frequents low bushy areas near the woodland edge, whilst another, the Striped Policeman, flies in and out of the high branches of sunlit trees. The Evening Brown haunts the lower layers of shaded woods, and in a few places at higher elevations on Mahé the large and showy King of Seychelles is occasionally seen passing rapidly through the trees. The latter species could be confused with the large migratory species of butterfly, the Diadem and African Monarch, which appear during the northwest monsoon, or with the more cosmopolitan Painted Lady.

Migratory hawk-moths can be attracted to outside lights on cloudy nights during the northwest monsoon and the two most regularly seen are the Convolvulus and Oleander Hawk-moths, although there are occasional records of other species.

### Damselflies & dragonflies (Odonata)

*(pages 134–147 – including a key to the species on pages 134–137)*

Damselflies are smaller and less robust than dragonflies. They have very slender bodies, the eyes are well separated and the narrow wings are generally held closed over and parallel to the body when at rest. As with dragonflies, the males are more frequently seen and the females are often best identified by association with their respective males.

Six species of damselfly are known from the area. These include two endemic species: the Black Mountain Damsel and the Seychelles Stream Damsel, both of which remain widespread and relatively numerous in damp forested areas on the three larger islands. Three of the non-endemic species are more common in the lowlands, although one, the Pygmy Wisp is rather localised. The remaining species, the Oceanic Fineliner of Mahé and Silhouette, was only rediscovered in 1997, having been thought to have been extinct since 1909.

Dragonflies are generally larger than damselflies. They have more robust bodies and are stronger fliers, the eyes touch and the wings are held outstretched from the body when at rest. A total of 13 species of dragonfly has been recorded from Seychelles. These include two endemic species and one species represented by an endemic subspecies.

The majority of the dragonfly species are widespread, and the populations of some strongly migratory species, such as Vagrant Emperor and Wandering Glider, are probably regularly augmented by influxes of migrants. These widespread dragonflies are mostly found in the lowlands, whereas the two endemic species, the Seychelles Cascader and Seychelles Duskhawker, are more restricted to higher elevations above 200m (650ft). Both are confined to the larger islands although the Seychelles Duskhawker is very rarely seen, partly because of its crepuscular habits. The Seychelles Skimmer is an endemic subspecies and is more widespread, being found commonly in both hill and lowland areas. The distinction between hill and lowland areas on Mahé and Praslin is real, but the distribution of species may simply be a reflection of the fact that the lowlands are more cultivated and the highland areas are wilder. Indeed, on Silhouette many 'highland' species are found at sea-level where there is very little cultivated land and the natural habitat has endured for many years.

## Other conspicuous invertebrates

This section of the book aims to enable the casual observer to identify those larger invertebrate species which are either very widespread and are therefore likely to be encountered, or are particularly striking in their appearance as to draw attention to themselves. Thus, we have included an ant-lion, two beetles, one cockroach, one bee, two wasps, the giant millipede, the giant centipede, the whip scorpion, two large spiders, eleven species of crab, two slugs and two snails. For further help on invertebrate identification, readers should consult the 'further reading' section, and refer to identification papers published regularly in *Phelsuma*, the annual journal of the Nature Protection Trust of Seychelles (NPTS).



It should be noted that many of these species do not have a common name. To aid the reader, however, we have added a descriptive name in these cases and these are shown with a '+' adjacent to the 'common' name given in the following species accounts.

BUTTERFLIES

## m Painted Lady 1

*Vanessa cardui*

Wingspan: 50 mm/2 in

A fairly large migrant butterfly that regularly appears during the northwest monsoon (December–April) following periods of NW winds. Generally scarce and most apparent on smaller offshore islands where it is often attracted to cultivated flowers in gardens.

**IDENTIFICATION:** A fairly large strong-flying butterfly, smaller and with a more complex pattern on the upperwing than African Monarch and female Diadem. **DESCRIPTION:** The complex upperwing pattern appears pink-brown in flight, and consists of black tips to the forewing with small white spots near the apex, and a yellowish-brown base bearing irregular black and brown markings. The underside is paler and has an irregular pattern of whitish and pale brown patches. There is a yellow-brown basal area on the forewing and five small dark-centred eye-spots near the rear edge of the hindwing. **BEHAVIOUR:** Flies rapidly and strongly, regularly coming to nectar on ornamental flowers. Often perches in sunny spots where it slowly opens and closes its wings, but closes its wings when alarmed and when in the shade.

## m African Monarch 2

*Danaus chrysippus*

Wingspan: 100 mm/4 in

A large, strong-flying migrant butterfly that can appear together with the very similar Diadem following strong NW winds and rain during the northwest monsoon (November–April). The bright orange-brown colour serves as a warning that this species is poisonous to potential predators. The very similar female Diadem is not poisonous but mimics this species in order to avoid predation. Often appears to be heavily outnumbered by Diadem in Seychelles.

**IDENTIFICATION:** Very similar to the slightly smaller female Diadem. Variation in both species means that positive identification of some individuals is impossible. The African Monarch tends to have three black spots towards the centre of the hindwing, which show on both the upper and undersides. Female Diadems typically have just one black spot, which is not always visible on the underside. **DESCRIPTION:** Usually rich orange-brown above, with dark blackish-brown tips to forewing and a variable series of large white spots across the corner of the wing tip. The fringes of the hindwing have an irregular dark border, and the forewing may show a pronounced pattern of black veins on both surfaces. There are normally three small black spots towards the centre of the hindwing, which show on both surfaces. **BEHAVIOUR:** The strong, rapid flight is distinctive, but also glides high on broad, outstretched wings. Recently arrived migrants often spend much time resting on the ground with their wings closed, but when nectaring on flowers often open their wings.

## m Diadem 3

*Hypolimnas misippus*

Wingspan: 80 mm/3 in

A large, strong-flying migrant butterfly that can appear in large numbers, together with the African Monarch, following strong NW winds and rain during the northwest monsoon (November–April).

**IDENTIFICATION:** Males are distinctive, their bold black upperwings bearing three rounded white patches. Females are very similar to female African Monarchs and may not always be safely distinguishable. **DESCRIPTION:** The male **(3m)** is largely black above and bears large, rounded, white patches, one on each wing, and a smaller white patch near the tip of the forewing. The underside is paler and browner and has pronounced darker veins. The female **(3f)** is typically orange-brown above with dark blackish-brown tips to the forewing and a variable series of large white spots across the corner of the wing tip. The fringes of the hindwing bear a regular dark border, and the upperwing may show a pronounced pattern of black veins on both surfaces, although this can be absent in some individuals. There is normally one small black spot towards the centre of the hindwing, which does not always show through to the lower surface of the wing. **BEHAVIOUR:** Much as African Monarch. Migrants sometimes breed successfully under suitable conditions.



life-size

1
2
3f
3f
3m
3m

## Evening Brown

1

Wingspan: 60 mm/2·3 in

*Melanitis leda*

A fairly large brown butterfly of rather variable appearance that flies low through the shaded understorey of woods, and is at least partially crepuscular. It is highly cryptic when perched. This species is widespread on the larger islands, occurring in wooded habitats from the coast up into the mountains, and may also migrate to smaller offshore islands where it is normally absent.

**IDENTIFICATION:** The only large brown butterfly that habitually flies through shaded lower layers of woods or along woodland paths at dusk. Could be confused with larger day-flying moths, but has a characteristic 'looping' flight on broad, blunt wings, and a rougher trailing edge to the hindwing. **DESCRIPTION:** A large, mostly brown butterfly which shows considerable variation in coloration. Typical individuals are dark brown on both surfaces, although the upperwing is usually a warmer brown and typically bears a large, white-centred, black eye-spot near the tip. There are often four smaller white-centred, black eye-spots near the rear edge of the hindwing. Some individuals can show a rich orange-brown centre to the upperside of the forewing, which may show through to the underside, while other individuals appear almost blackish. **BEHAVIOUR:** Flies low with a looping moth-like flight through the shade of the lower layers of woods. Most active at dawn and dusk, but may also fly during the day, particularly during overcast conditions or if flushed from its hiding place. Resembles a dead leaf when perched with wings closed, often tilting away from the observer and becoming remarkably hard to spot.

## Striped Policeman

2

Wingspan: 40 mm/1·6 in

*Coeliades forestan*

A local resident that occurs throughout the year along the sunlit edges of wooded habitats, flying rapidly high up around flowering and fruiting trees.

**IDENTIFICATION:** The rapid flight high around sunlit trees is distinctive, together with the flash of a bold white vertical bar on the dark chocolate-brown hind-wing. Black and white stripes on the abdomen are diagnostic, but these can normally only be seen when perched. **DESCRIPTION:** A stocky, mid-sized butterfly, which is largely dark brown in colour, but shows a bold white vertical stripe on both surfaces of the hind-wing. The plump, short abdomen is orange at both ends and has a series of broad white and thinner black stripes. The head and first joint of the legs are also orange. The eyes are large and dark. **BEHAVIOUR:** Flies quickly around trees with rapid wing-beats, only rarely and briefly settling to rest or to sip sugars from flowers or ripe fruit. It is active throughout the day, but particularly early in the morning and again towards dusk, roosting in the evening by hanging upside-down on the underside of leaves. It generally rests with its wings closed.

## Olive-haired Swift

3

Wingspan: 35 mm/1·4 in

*Borbo borbonica*

A small, mid-brown skipper butterfly that is widespread but local, favouring sunlit patches of vegetation close to the edge of wooded habitats. It can be found throughout the year but is most numerous during and after wetter months.

**IDENTIFICATION:** The only widespread small, brown skipper butterfly in Seychelles. It is readily identified by its small size, and the presence of small, translucent white spots on the forewing. The wings are otherwise all mid- to dark brown. This species lacks the bold white stripe on the hindwing of Striped Policeman. **DESCRIPTION:** The body is short and broad with an even broader head. The wings are mid- to dark brown above and below, with up to seven translucent whitish spots on the forewing, including two or three larger, more obvious, spots and several smaller, less distinct ones. Some males also show similar tiny pale spots on the hindwing. **BEHAVIOUR:** A typical skipper butterfly, which flies with rapid whirring wing-beats and which rests on low sunlit vegetation with its forewings and hindwings held at different angles to each other.



life-size

## E King of Seychelles 4

*Euploea mitra*

Wingspan: 60 mm/2·3 in

A large endemic butterfly that is currently thought to be restricted to areas of forest at higher elevations on Mahé, where it is rarely seen and may be endangered. There are old records from high, forested areas on Silhouette.

**IDENTIFICATION:** A large, strong-flying butterfly, which is easily separated from the similar-sized African Monarch and Diadem butterflies *(page 126)*, by its largely blackish-brown wings, which bear a broad, pale yellowish band towards the rear edge of the hindwing and often a bold, whitish band across the front corner of the forewing. **DESCRIPTION:** The wings are mostly blackish-brown, with the distinctive bold patterning described above. The body is quite long and is dark in colour. **BEHAVIOUR:** Flies rapidly and strongly, high through the trees, usually affording only brief glimpses. Recent records have come from the Mission/Salazie area of Mahé, and the species requires further study to determine its current status.

## Long-tailed Blue

1

Wingspan: 32 mm/1·3 in

*Lampides boeticus*

A large blue butterfly that can be locally common in areas with scattered bushes and flowering leguminous plants, particularly on La Digue. Its similarity to Lang's Short-tailed Blue may mask its true status elsewhere. Adults can be found flying throughout most of the year.

**IDENTIFICATION:** Obviously larger and generally higher-flying than African Grass Blue. It is very similar to Lang's Short-tailed Blue, but is larger, with longer 'tails' and has larger black eye-spots bearing a broader orange fringe on the underwings; it also usually shows a pronounced pale stripe towards the rear edge of the underside of the hindwing.
**DESCRIPTION:** A large blue butterfly. Males have violet-blue wings above with blackish fringes and two dark black spots on the hindwing close to the long 'tails'. Females are mostly brown above, with a variable area of violet-blue restricted to the central areas of the upperwing. The underside is pale brown with many thin wavy pale bands, and usually one broader, whiter stripe near the rear edge of the hindwing. There are two pronounced black eye-spots at the base of the long black 'tail', which are bordered broadly with orange.
**BEHAVIOUR:** Flies rapidly and strongly often quite high over bushes and shrubs. It often pauses to feed on flowers, particularly those of the legume family. It is most approachable at rest, when the wings are normally held closed.

## Lang's Short-tailed Blue

2

Wingspan: 25 mm/1 in

*Syntarucus pirithous*

A mid-sized blue butterfly, the true status of which is masked by its similarity to the larger and locally more numerous Long-tailed Blue.

**IDENTIFICATION:** Intermediate in size between African Grass Blue and Long-tailed Blue, but more easily confused with the latter, as it also has fast, high-flying habits and a similar pattern on the underside. Best separated from Long-tailed Blue by the smaller black eye-spots on the underside that lack a broad orange fringe, and the lack of a pronounced whitish stripe towards the rear edge of the underside of the hindwing.
**DESCRIPTION:** A mid-sized blue butterfly. Males have violet-blue wings above with darker fringes and two small dark black spots on the hindwing close to the longish 'tails'. Females are mostly grey-brown above, with a small area of violet-blue restricted to the central areas of the upperwing. The underside is brown with many thin, wavy pale bands, but normally lacks a broader whiter stripe on the hind-wing. There are two pronounced black eye-spots at the base of the long black 'tail', which have a thin orange border.
**BEHAVIOUR:** Much as Long-tailed Blue, but flight less powerful.

## African Grass Blue

3

Wingspan: 20 mm/0·8 in

*Zizeeria knysa*

A widespread and often abundant small blue butterfly that flits restlessly low over open vegetation, particularly short-mown lawns. Most numerous in the wetter northwest monsoon, though flies year-round.

**IDENTIFICATION:** Can be separated with care from the other blue butterflies. It is distinctly smaller than Lang's Short-tailed Blue and the Long-tailed Blue, and lacks the 'tails' and underwing eye-spots of these species. It is most similar to Tiny Grass Blue, but is larger, has a browner underside and the male is a darker, more violet blue above.
**DESCRIPTION:** A very small blue butterfly. Males have violet-blue wings above with dark brown fringes, whereas females are mid-brown above. In both sexes the underside is pale grey-brown with small black spots fringed white and there are no larger eye-spots.
**BEHAVIOUR:** Flits restlessly over grass lawns and other low vegetation, often in quite large numbers. It rarely settles, but pauses briefly to sip nectar from flowers, when the wings are normally held closed.



2× life size

## Tiny Grass Blue

4

*Zizula hylax*

Wingspan: 16mm/0·6in

A widespread, tiny blue butterfly that flits low over open areas of taller grass. More common in the wetter northwest monsoon.

**IDENTIFICATION:** A truly tiny blue butterfly that occurs in moist, grassy lowland habitats. It tends to fly singly and is always outnumbered by the very similar but distinctly larger African Grass Blue. However, Tiny Grass Blue has a silvery-white underside and the male is a paler powder blue above.

**DESCRIPTION:** A tiny blue butterfly. Males have pale powder blue wings above, with darker fringes. The underside is pale silvery-white with small black spots and there are no larger eye-spots.

**BEHAVIOUR:** Similar to African Grass Blue, but rests more often and prefers taller grass. The wings are generally closed when perched.

MOTHS

The Seychelles are home to a large number of moths, both resident and migrant. An identification guide to all these species is beyond the scope of this book. However, shown here are two large obvious hawk-moths that are often attracted to lights

## m Oleander Hawk-moth 1

*Daphnis nerii*

Wingspan: 95 mm/3·7 in

A large migrant hawk-moth that occurs during the northwest monsoon (December–April) following periods of NW winds. It is most often observed when attracted to lights at night, though is generally less numerous than the Convolvulus Hawk-moth.

**IDENTIFICATION:** A large, strong-flying nocturnal moth, that is best separated from the similar Convolvulus Hawk-moth, by its smaller size, more pronounced pattern of white, brown and olive markings on the forewings and the absence of broad pink bars on the abdomen. Close views reveal two pale patches with a dark central spot on either side of the body at the base of the forewings, which are diagnostic.

**DESCRIPTION:** A large hawk-moth with a pronounced and complex pattern of whitish, pale brown and olive markings on the forewings and more diffuse whitish and brown markings on the small hindwings. There is a small pale area bearing a central dark spot at the base of each forewing. The thorax is broad and brownish, with darker olive-brown triangles on the 'shoulders'. The broad abdomen has diffuse alternating bands of darker and paler brown, often with a whitish band near the front.

**BEHAVIOUR:** Much as Convolvulus Hawk-moth. Flies rapidly around night-flowering plants, and is readily attracted to lights at night. It is generally scarce, and breeding has not been recorded in Seychelles.

## m Convolvulus Hawk-moth 2

*Agrius convolvuli*

Wingspan: 105 mm/4·1 in

A large migrant hawk-moth that regularly appears during the northwest monsoon (December–April) following periods of NW winds. Most often observed when attracted to lights at night. Attempts to breed in some years.

**IDENTIFICATION:** A large strong-flying nocturnal moth, that is best separated from the similar Oleander Hawk-moth, by its larger size, more diffuse pattern of grey, brown and olive markings on the wings, and the presence of broad pink bars on the abdomen.

**DESCRIPTION:** A large hawk-moth with a rather complex and nondescript pattern of grey, brown and olive markings on the forewings and thicker, more even, wavy bands of grey and olive-brown on the small hindwings. The broad head and thorax are brownish, and there is a paler central area on the back. The abdomen is broad and distinctively striped pink and blackish-brown. The pink stripes are broader towards the head, becoming thinner and less obvious towards the tail, which is largely blackish-brown.

**BEHAVIOUR:** Flies rapidly and strongly, buzzing around night-flowering plants. It is readily attracted to lights at night, when it flies in diminishing circles or perches close to the light source. They are more numerous in some years than others, and sometimes breed. The large green caterpillars have a prominent black-tipped reddish horn at one end and feed readily on the leaves of native trees, such as *Cordia subcordata* and *Ochrosia oppositifolia*, before pupating in the cover of leaf-litter.

1

2

# LIFE SIZE KEY TO THE DRAGONFLIES AND DAMSELFLIES OF SEYCHELLES

### Damselfly or Dragonfly?

DAMSELFLY:
Wings held over and parallel to the body when resting. Eyes separated

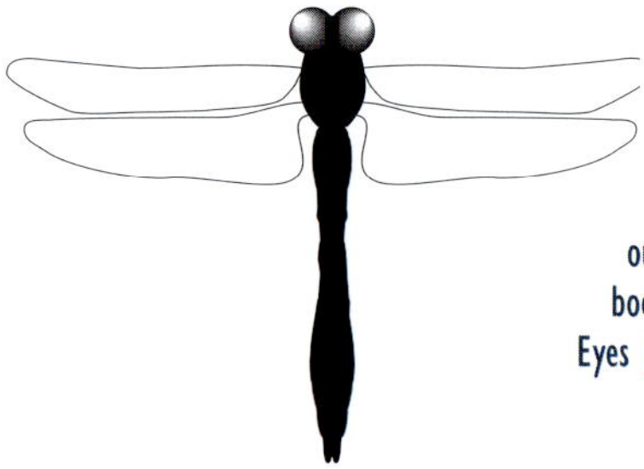

DRAGONFLY:
Wings held outstretched from body when resting. Eyes joined at centre

## DAMSELFLIES (Pti Sigal)

Location – COASTAL or HIGHLAND?

Key: = common
= endemic species

### COASTAL

**Pygmy Wisp**
PLATE 52

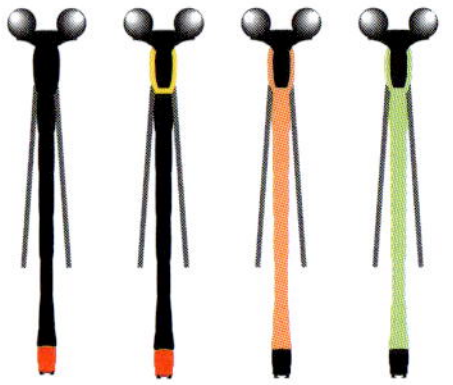


**Marsh Bluetail**
PLATE 52

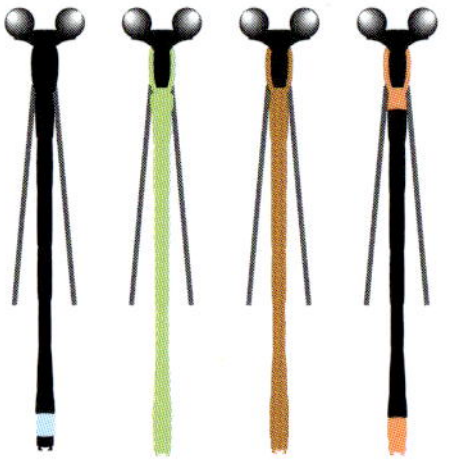


**Common Pond Damsel**
PLATE 51

**Oceanic Fineliner**
PLATE 52

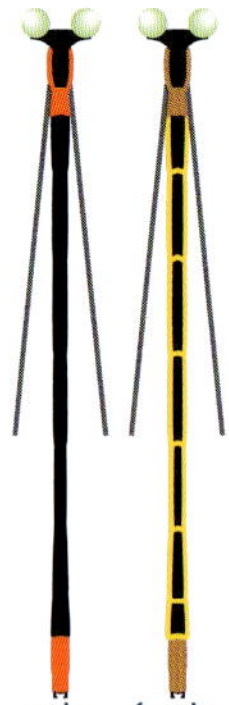


### HIGHLAND

**Black Mountain Damsel**
PLATE 51

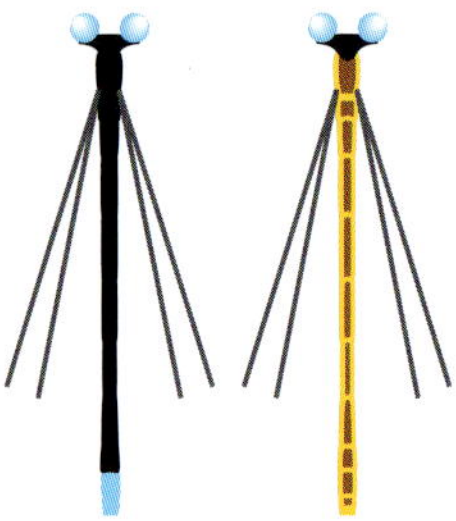


**Seychelles Stream Damsel**
PLATE 51

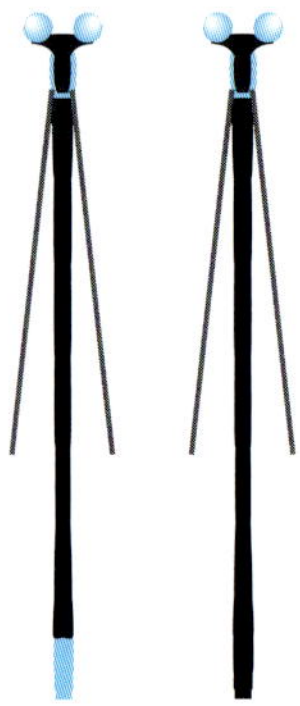

**DRAGONFLIES** (Sigal)
Main abdomen colour

LOCATION –
COASTAL or HIGHLAND?

When do you see them? –
ANY TIME or DAWN & DUSK ONLY?

Key: = commonly seen
= endemic species

Key: = coastal species
= highland species

Key: = all day
= dawn and dusk only

## DRAGONFLIES WITH RED or BROWN ABDOMENS

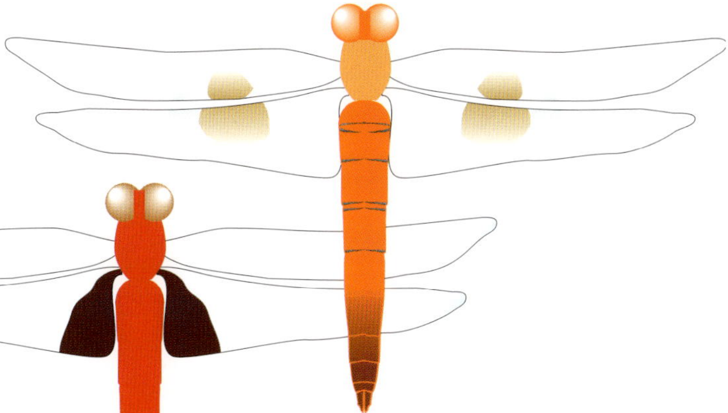

Twister
PLATE 54

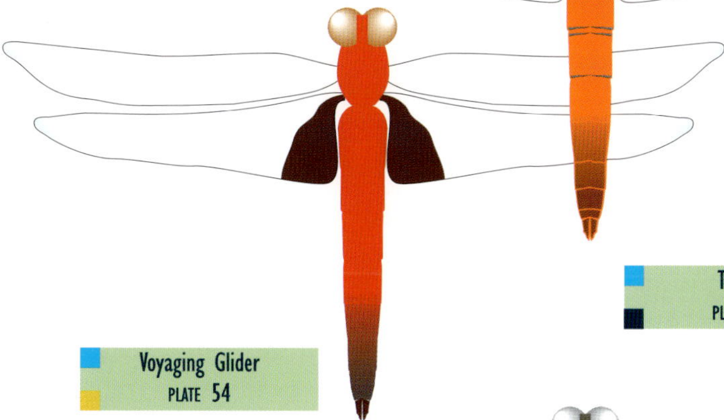

Voyaging Glider
PLATE 54

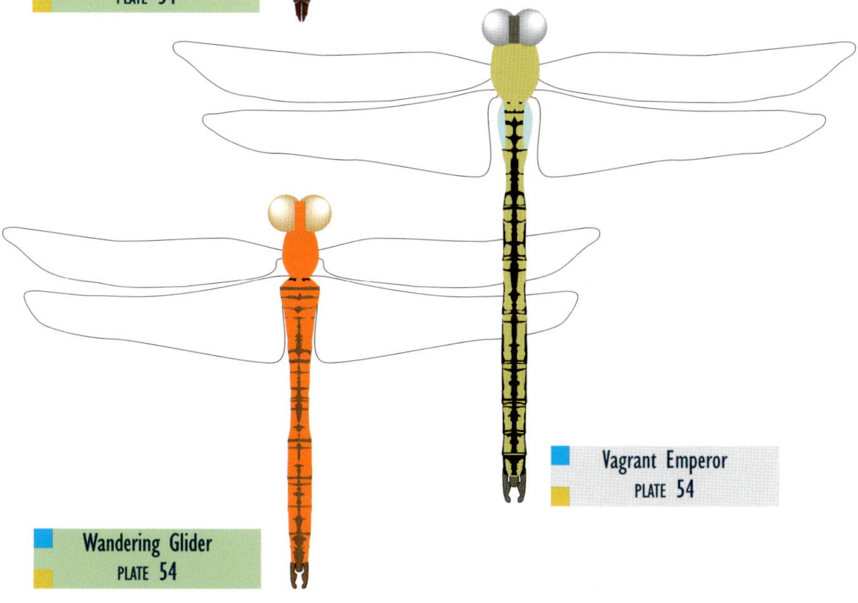

Vagrant Emperor
PLATE 54

Wandering Glider
PLATE 54

**DRAGONFLIES** (Sigal)
What colour is the abdomen mainly?

LOCATION –
COASTAL or HIGHLAND?

When do you see them? –
ANY TIME or DAWN & DUSK ONLY?

Key: = commonly seen
= endemic species

Key: = coastal species
= highland species

Key: = all day
= dawn and dusk only

## DRAGONFLIES WITH BLUE ABDOMENS

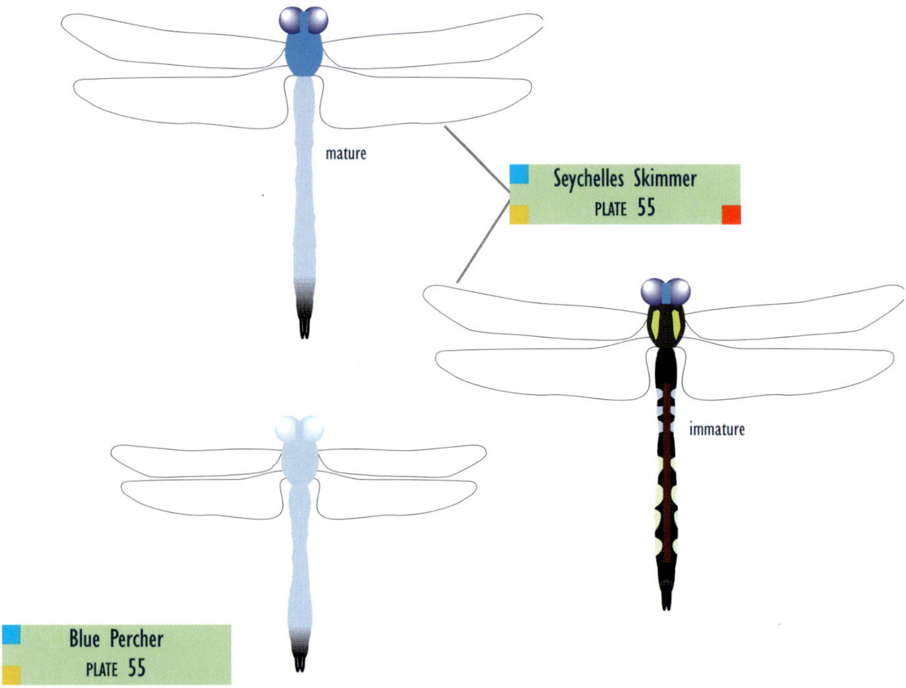


## DRAGONFLIES WITH GREEN ABDOMENS

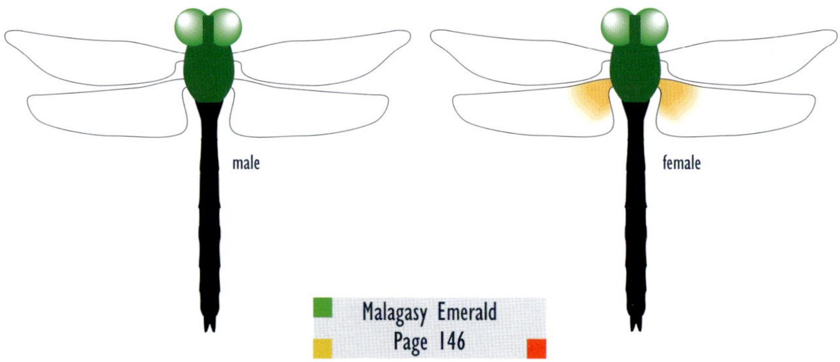

# DRAGONFLIES WITH BLACK ABDOMENS

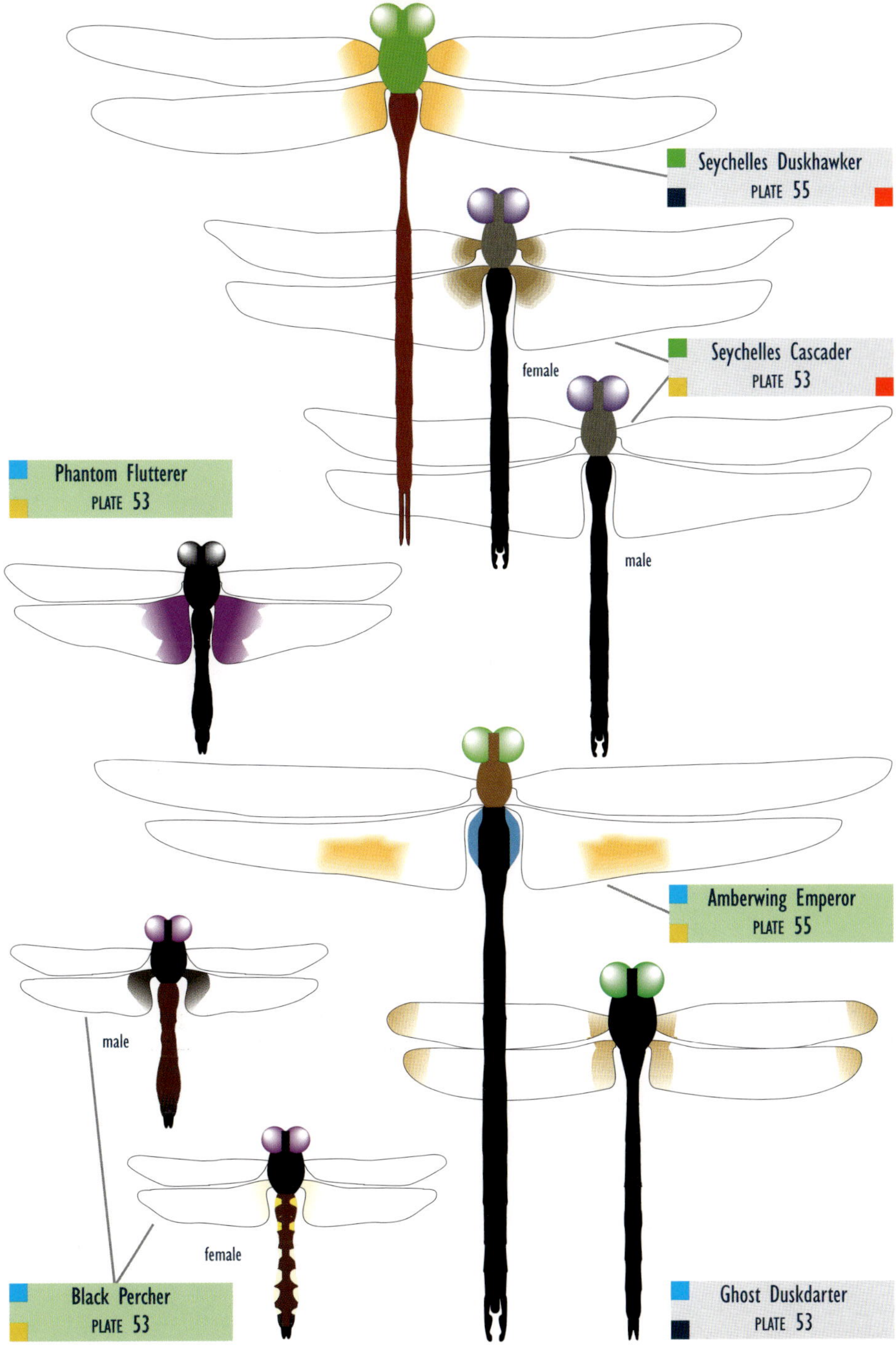

DAMSELFLIES Pti Sigal

## E Black Mountain Damsel 1

*Allolestes maclachlanii*

Length: 32 mm/1·3 in Wingspan: 42 mm/1·7 in

Common in forested areas close to streams in the hills of the main islands. Most common at higher altitudes on Mahé and Praslin away from cultivated (developed) areas. Also occurs lower down in appropriate habitat on Silhouette.

**IDENTIFICATION:** A very long-bodied, slim, dark damsel with a weak, fluttery flight and a pronounced pale blue band near the tip of the abdomen. It usually rests with wings half-spread. This species is extremely similar to Seychelles Stream Damsel, from which males are best separated by the size and shape of the blue tip to the abdomen and relatively short body. In Black Mountain Damsel, this blue marking is restricted to the top surface of the last two-and-a-half abdominal segments, which are otherwise black. This species is found in a very different habitat from the Marsh Bluetail, which has a shorter, thicker abdomen, more powerful flight and is an open-country lowland species. **DESCRIPTION:** Body of male **(1m)** is largely black, but the eyes are blue, the face and head have blue areas and there are blue markings on the thorax and abdomen. The female **(1f)** is brown, with yellow banding at the segments of the abdomen. The legs are pale brown and bear long, slender spines. **BEHAVIOUR:** Flutters weakly low to the ground in damp, shady undergrowth of forests close to mountain streams, perching on the ground or low vegetation. Often found away from water. **BREEDING:** Egg-laying probably occurs year-round.

## E Seychelles Stream Damsel 2

*Leptocnemis cyanops*

(formerly Blue-eyed Mountain Damsel)

Length: 44 mm/1·7 in Wingspan: 48 mm/1·9 in

Common in shady forested areas close to streams in the hills of the main islands. Most common at higher altitudes on Mahé and Praslin away from cultivated (developed) areas, but also occurs lower down in suitable habitat on Silhouette.

**IDENTIFICATION:** Very similar to Black Mountain Damsel, but has an even longer, slim, dark body and a less pronounced pale blue band near the tip of the abdomen. The blue patch on the male abdomen is restricted to the last but one segment, which is entirely bright blue, and to the sides of the final segment, which has a black top surface. **DESCRIPTION:** The abdomen and thorax of male **(2m)** are mostly black, although the abdomen is paler along the sides, the eyes are bright blue, the face and head bear bright blue markings and there are bright blue markings on the thorax and abdomen. The legs are pale brown and bear long, slender spines. Females **(2f)** are similar to males but the abdomen is entirely black above. **BEHAVIOUR:** Has a weak, fluttery flight and keeps low to the ground as it patrols shady parts of damp forest undergrowth close to mountain streams, perching on the ground or on low vegetation. Likes to hang off the tips of leaves and grasses and is hardly ever seen away from streamsides. **BREEDING:** Egg-laying probably occurs year-round. Ovipositing occurs as the insects fly in tandem.

## Common Pond Damsel 3

*Ceriagrion glabrum*

(formerly Orange Damselfly or Red Damsel)

Length: 40 mm/1·6 in Wingspan: 50 mm/2·0 in

A common and widespread species, also found in Africa. It favours sunlit wetland habitats in plateau areas and lower reaches of hills.

**IDENTIFICATION:** A large, sturdy damselfly, identified by its uniformly bright orange (male) or orange-brown (female) abdomen, which is substantially broader than in other damselflies.
**DESCRIPTION:** Males **(3m)** are easily distinguished by their bright reddish-orange abdomen. Females **(3f)** are variably dull brown to orange-red, with a browner terminal section of the abdomen. The thorax sometimes has a greenish tinge. Teneral specimens are very pale creamy brown and have the characteristically weak flight of all freshly emerged damsels. **BEHAVIOUR:** Males patrol clear, sunlit areas, including paths, in plateau areas. It is less numerous and more strictly confined to wetland sites during the drier southeast monsoon. **BREEDING:** Egg-laying occurs throughout the year but is less frequent during dry spells. Ovipositing occurs as the insects fly in tandem.



2× life-size

1m
1f
2f
2m
3m
3f

## Marsh Bluetail

1

*Ischnura senegalensis*

(formerly African Blue-tailed Damselfly or African Damsel)

Length: 28 mm/1·1 in Wingspan: 30 mm/1·2 in

A widespread species, which occurs in Africa and Asia. Locally common in wetland habitats in plateau areas with still or slow-moving water.

**IDENTIFICATION:** Slimmer and slightly shorter (*ca.* 3/4 length) than Common Pond Damsel *(page 138)*, and easily identified by its largely dark body with a distinct pale blue band near the tip of the abdomen. It is superficially similar to the endemic mountain and stream damsels *(page 138)*, but has a much shorter, relatively thicker abdomen and a stronger flight. It also has different habitat preferences. **DESCRIPTION:** Males **(1m)** are generally black but with a prominent pale blue band near the tip of the abdomen, pale blue dots behind the eyes and a rich bluish-green thorax. There are three colour variants of the female: some show the same basic pattern as the male but with greener sides **(1i)**; some are dull orange-brown; and some, probably a teneral variant, have a bright orange head and thorax and orange on the first and last segments of the abdomen **(1iii)**. **BEHAVIOUR:** Males patrol open sunlit areas in the lowlands, especially along streams and the edges of marshes. They frequently perch on vegetation and habitually pump their abdomens slowly up and down. **BREEDING:** Pairs have been seen mating in May on La Digue; probably breeds mostly during the wetter northwest

## Pygmy Wisp

2

*Agriocnemis pygmaea*

(formerly Pygmy Firetaill)

Length: 23 mm/0·9 in Wingspan: 23 mm/0·9 in

A widespread species, also found in Asia. It occurs locally in wetland habitats with still or slow-moving water in plateau areas.

**IDENTIFICATION:** A very small and slight species, only a half to two-thirds the length of the other two common plateau damsels. **DESCRIPTION:** Males **(2m)** are generally all blackish with a fiery orange-red tip to the abdomen, and lack obvious markings on the wings. As they age, the orange tip to the abdomen becomes reduced or even absent and a blue-white pruinesence develops on the normally black thorax. There are three colour variants of the female: as male but with a yellower thorax **(2i)**; a red form with a black-tipped, brick-red abdomen **(2ii)**; a (probably teneral) form with a black-tipped green abdomen **(2iii)**. **BEHAVIOUR:** Flies very weakly, spending much time resting on stems of emergent vegetation and nearby long grass, although it can turn up sporadically on offshore islands where it is otherwise absent. **BREEDING:** Poorly known in Seychelles.

## Oceanic Fineliner

3

*Teinobasis alluaudi*

(formerly Slender Damselfly)

Length: 44 mm/1·7 in Wingspan: 46 mm/1·8 in

A vulnerable species that was long thought to be extinct until its rediscovery on Mahé in 1997 and Silhouette in 1998, having last been recorded from Mahé in 1909 and Silhouette in 1908. Although previously recorded from both lowland and montane wetlands, all recent records have come from shaded ponds or streams in the lowlands.

**IDENTIFICATION:** The very long and slender body (abdomen length 30 mm in males and 34 mm in females) is distinctively coloured (see below) and readily separates this species from all other damselflies in Seychelles. **DESCRIPTION:** Males **(3m)** are all black with a rich red tip to the abdomen, red thorax and adjacent front part of the abdomen, and lack obvious markings on the wings. The eyes are greenish, paler on the underside. Females are largely yellowish but also have black and red markings. **BEHAVIOUR:** Flies over partly shaded stretches of streams, but also rests on stems of emergent vegetation. They have been observed to hover with the abdomen hanging vertically whilst removing prey from spiders' webs. **BREEDING:** Poorly known.



2× life-size

1iii
1m
1i

2i
2ii
2m
2iii

3

DRAGONFLIES Sigal

## Ghost Duskdarter

1

*Zyxomma petiolatum*

(formerly Ghost Skimmer or Bloated Skimmer)

Length: 50 mm/2·0 in Wingspan: 66 mm/2·6 in

A local, crepuscular species also found in Asia. In Seychelles it is restricted to shaded wetland sites in plateau areas, normally close to the coast.

**IDENTIFICATION:** The rapid, darting flight low over water is a distinctive feature of this species. It is very rarely seen perched, so is hard to observe closely. The general impression is of a small, thin, all-dark dragonfly with large pale green eyes. **DESCRIPTION:** The abdomen is blackish (or dull brown in females) and very thin with a slightly swollen base. The thorax is dark with an ochraceous hue, and the face is dark with prominent large apple-green eyes. In males, the wings have brown tips and bases, whereas in females the wings just have brown bases. **BEHAVIOUR:** Males patrol a regular beat over open pools, flying very rapidly low over the water and constantly changing direction, briefly hovering in one spot, before darting off again. They generally fly just after dawn and again between 16.00 and 18.30 hrs, although may fly at other times on heavily overcast days. **BREEDING:** Females lay eggs alone, sometimes with a male in attendance. More common in the wetter northwest monsoon, although odd individuals are seen during the drier southeast monsoon.

## E Seychelles Cascader

2

*Zygonyx luctifera*

(formerly Steely Skimmer)

Length: 51 mm/2 in Wingspan: 80 mm/3·1 in

A locally common endemic species, normally found in association with waterfalls, cascades and fast-flowing streams at higher altitudes (>200 m) in wooded areas on Mahé and Praslin, but occurring at lower elevations on Silhouette.

**IDENTIFICATION:** A mid-sized, black dragonfly with a slim, slightly bulbous-tipped abdomen, unmistakable if seen well. The Phantom Flutterer is also largely black, but has a shorter, squatter abdomen, pronounced dark wing patches close to the body, a fluttery flight on short broad wings and is restricted to the lowlands. **DESCRIPTION:** The body of the male **(2m)** is almost entirely black with a glossy blue-green sheen. The thorax is browner but retains the glossy sheen, whilst the face is paler. The eyes are metallic black, with a blue gloss. The wings lack obvious markings. Females **(2f)** differ in having a paler base to the abdomen and, more noticeably, a pronounced triangle of golden-amber at the base of the hindwings close to the body, and a similar but smaller patch at the base of the forewings. **BEHAVIOUR:** Males are very active, rapidly patrolling beats with strong, direct flight both low over the ground and high over trees. Females also fly high above the canopy. **BREEDING:** Females oviposit alone, splashing their abdomens onto the surface of the water.

## Phantom Flutterer

3

*Rhyothemis semihylina*

(formerly Brunette Chaser)

Length: 28 mm/1·1 in Wingspan: 56 mm/2·2 in

A widespread species, also found in Africa, which is locally common in wetland habitats in plateau areas with still or slow-flowing water, particularly close to the shore. It is at least a partial migrant, with individuals appearing after strong winds on islands where it is not normally resident.

**IDENTIFICATION:** Similar in size to the Blue Percher *(page 146)*, but appears squatter, and has an obviously fluttery flight. The body is entirely blackish and the short wings have bold dark patches close to the body. **DESCRIPTION:** Males and females are similar, the short abdomen, thorax, face and eyes being entirely black. The broad hindwings show two pronounced large blackish patches adjacent to the body, which reflect iridescent purple tints in sunlight. **BEHAVIOUR:** Flutters low over wetlands, often moving sideways rather than forwards in the breeze. **BREEDING:** Probably breeds mostly during wetter northwest monsoon but active at wetland sites throughout the year.



life-size

## Black Percher (4)

*Diplacodes lefebvrii*
(formerly Ebony Skimmer)

**Length:** 28 mm/1·1 in **Wingspan:** 45 mm/1·8 in

A rather local species, also found in Africa and Asia. It occurs in plateau areas and is often seen well away from water. It is a strong flyer, which may appear on islands where it is not normally resident, following onshore winds.

**IDENTIFICATION:** A small, blackish-brown dragonfly, most easily confused with similar-sized Phantom Flutterer but has much smaller dark patches only at the base of the hindwing.
**DESCRIPTION:** Males **(4m)** appear rather drab, with a broad dark brown abdomen tipped black, a blackish thorax and dark purple eyes. There are distinct black patches on the hindwing only, close to the body. Females **(4f)** are similar to males but are duller and paler. At rest the abdomen is characteristically held in a down-pointing position. **BEHAVIOUR:** Males make hunting sallies from a perch, often well away from wetlands. **BREEDING:** Poorly known in Seychelles.

## Voyaging Glider

1

Length: 50 mm/2·0 in Wingspan: 95 mm/3·7 in

*Tramea limbata*

(formerly Scarlet Chaser *Trapezostigma limbatum continentale*)

A common and widespread species, also found in Africa. It hawks over wetlands and adjacent open areas both in plateau and hill areas. Adults are present throughout the year, although are more numerous and less confined to wetlands during the northwest monsoon.

**IDENTIFICATION:** The common medium-sized red dragonfly in Seychelles, easily separated from other species by its broad, bright scarlet abdomen. **DESCRIPTION:** Males have a broad, bright scarlet abdomen, which is generally darker (blackish) at the tip. Rusty-red patches may be prominent in the wings close to the body, though these vary in both extent and colour. In flight these patches give the appearance of saddle-bags, and the abdomen is sometimes angled slightly downwards. Females have the same basic colour pattern as males but are duller and browner. The wings are broad and mid-length. Some individuals show distinct opaque white tips to all four wings, though this may be a feature suggesting recent emergence, as it fades over the course of a few days. The eye colour is variably red to blackish with a bluish sheen. **BEHAVIOUR:** Males patrol a regular small beat relentlessly throughout the day, often returning to a favoured perch. **BREEDING:** Breeds throughout the year, females lay eggs by tail-dipping, either alone or in tandem with a male.

## m Vagrant Emperor

2

Length: 63 mm/2·5 in Wingspan: 94 mm/3·7 in

*Anax ephippiger*

(formerly Blue-hipped Hawker *Hemianax ephippiger*)

A strongly migratory species that also occurs in Africa and Asia. It appears in some numbers over wetlands and associated open parts of plateau areas following periods of strong winds and rain during the northwest monsoon, particularly from December–April.

**IDENTIFICATION:** A fairly large, strong-flying dragonfly characterised by a relatively short, slim and bulbous-tipped abdomen that is distinctly shorter than the hindwing. It is similar to Amberwing Emperor *(page 146)* but the blue base to the mid- to pale brown (not black) abdomen is restricted to the upper surface and is less pronounced. **DESCRIPTION:** The abdomen is mid-brownish-grey above with a pale blue band at the base, the remainder being indistinctly banded with pale blue and with a paler rufous tip. The underside is paler. The very large, grey-green eyes are set in a pale grey-green face. The wings often show a rufous tone in good light. **BEHAVIOUR:** A strong flyer that glides high on an irregular beat, often displaying a similar vertical wing wobble to Wandering Glider. It patrols a beat throughout the day over wetlands, but is also found in larger groups over open areas and is particularly active at dusk, when it flies with a more rapid darting flight low over beach-crest areas. **BREEDING:** Breeding status in Seychelles is unclear, numbers being swelled by migrants.

## Wandering Glider

3

Length: 50 mm/2·0 in Wingspan: 82 mm/3·2 in

*Pantala flavescens*

(formerly Globe Skimmer or Storm Chaser)

A very widespread and common pantropical species. Hunts in open areas, often well away from water and is commonly seen in towns. Most common in plateau areas but also ranges into the hills. Influxes occur following periods of strong winds and rain during the northwest monsoon, bringing it to islands where it may not be normally resident.

**IDENTIFICATION:** A mid-sized species, characterised by its long, pointed orange abdomen, and steady, high patrolling flight on strong, triangular-shaped hind wings. The similar-sized Voyaging Glider has a much deeper red abdomen. **DESCRIPTION:** The abdomen, thorax and head are all strikingly orange, some males showing a reddish tinge to the lower part of the abdomen. The eyes are yellow-orange. The wings are distinctly triangular in shape, and lack obvious markings. **BEHAVIOUR:** Flies strongly, patrolling a regular beat, the wings often quiver with a slight vertical wobble. **BREEDING:** Breeds opportunistically throughout the year, whenever water bodies appear. Often lays eggs in temporary pools such as roadside puddles. Females lay eggs by tail-dipping, either alone or in tandem with a male.



0·75× life-size

1

4

2

3

## Twister

4

*Tholymis tillarga*

(formerly Branded Skimmer)

| Length: 50 mm/2 in | Wingspan: 95 mm/3·7 in |
|---|---|

A widespread, crepuscular species, also found in Africa and Asia. It is locally common at lowland wetland sites, particularly close to the sea. Most numerous during the northwest monsoon.

**IDENTIFICATION:** A mid-sized dragonfly, characterised by a bright orange-red abdomen which tapers towards the tip, rapid darting flight low over the water and distinct pale white and fawn patches in both wings. The hindwing is very broad. **DESCRIPTION:** Males have a bright orange-red abdomen, which tapers and is slightly darker at the tip. The eyes are red and the thorax is yellower than the abdomen. The diagnostic features are the pronounced spots on the inner portion of the central part of each wing. Each spot is of two colours, darker and browner on the inner half and paler and whiter on the outer half; this is most apparent when they are flying at dusk. Females are similar to males but are duller and the abdomen is a browner-red. **BEHAVIOUR:** Males patrol erratic beats at dawn and again from late afternoon to dusk, flying fast and low over water and adjacent open areas, such as beaches. They dart about irregularly with rapid wing-beats and settle only rarely. When males meet over open water they engage in an ascending spiralling 'twisting' flight. During the middle part of the day males may be seen resting on reed stems, with just their wing spots showing on either side of the stem. **BREEDING:** Females oviposit at dusk, either alone, or with males in attendance. Breeding occurs mostly during the northwest monsoon.

## E Seychelles Duskhawker

1

*Gynacantha stylata*

(formerly Rusty Hawker)

Length: 72 mm/2·8 in Wingspan: 95 mm/3·7 in

A rare, poorly-known crepuscular dragonfly recorded from wooded hills above 200 m (650 ft) on Mahé, Praslin and Silhouette.

In flight in the dim light of late dusk this species appears only as a grey movement, quartering the ground less than 50 cm above the surface. It has been observed coming to house lights in search of insect prey. Males and females are very similar. The abdomen is black or very dark green with small, pale green markings, the thorax is bright green, and there are distinct amber patches, which increase with age, to the bases of both wings. The anal appendages are characteristically very long and slender.

## m? Malagasy Emerald

Length: 40 mm/1·6 in Wingspan: 55 mm/2·2 in

*Hemicordulia similis* (not illustrated)

(formerly Greenback Hawker)

A rare, poorly-known species recorded only once from mountains on Mahé. Possibly just a rare migrant to Seychelles.

A day-flying species, which patrols a repeated steady beat. Best identified by its brilliant metallic green thorax. The male lacks prominent wing-patches, whereas the female has a basal amber patch to the hindwing only. The only known record from Seychelles was in 1952.

## Amberwing Emperor

2

*Anax guttatus*

Length: 82 mm/1·5 in Wingspan: 112 mm/4·4 in

The common large emperor dragonfly of Seychelles, also found in Asia. It may be seen in small numbers throughout the year hawking over open wetlands and adjacent open areas. It is most common in plateau areas but also ranges into the hills.

**IDENTIFICATION:** A strong flying species, identified by its large size and blue-hipped appearance. This is due to the blue base to the otherwise blackish, long abdomen. The Vagrant Emperor *(page 144)* also has a pale blue base to the abdomen, but this is less pronounced, the abdomen is paler and browner, and is always shorter than the wings. **DESCRIPTION:** In flight, the abdomen of males appears blackish, with a distinct pale blue band at the base. The thorax is yellower and the eyes are green. There is an amber panel near the middle of the hindwing, though this is hard to see, except at close range in good light. Females are similar to males although are duller and browner. **BEHAVIOUR:** Males patrol regularly and repeatedly over open pools during the afternoon and towards dusk, but are less active during the morning. **BREEDING:** Breeds mostly during the wetter northwest monsoon, females sometimes laying eggs alone into the stems of aquatic plants.

## Blue Percher

3

*Diplacodes trivialis*

(formerly Skyblue Chaser/Skimmer)

Length: 33 mm/1·3 in Wingspan: 52 mm/2·0 in

A widespread species, also found in Asia. It occurs in wetland habitats in plateau areas with still or slow-flowing water, and in the hills.

**IDENTIFICATION:** A small, restless pale blue dragonfly, which frequently perches for short periods on wetland plants. It is easily confused with the blue form of Seychelles Skimmer but is smaller, shorter-winged, has a less powerful flight and the abdomen is often tilted upwards (not downwards) at rest. **DESCRIPTION:** Males and females are similar, having a pale powdery blue abdomen, thorax and eyes, and a paler face. The abdomen is pinched and flattened vertically with a fairly prominent keel and has a blackish tip often with prominent white anal appendages. Some females have pale greenish abdomens with bold black bars including a thick black terminal bar that is tipped white. **BEHAVIOUR:** Buzzes restlessly low over open water with rapid, direct flight, perching for short periods on nearby vegetation. It is less numerous and more confined to wetland sites during the drier southeast monsoon. **BREEDING:** Egg-laying occurs mostly during the wetter northwest monsoon, the females ovipositing by tail-dipping onto the surface of the water, either alone, or in tandem with a male.



0·75× life-size

## Seychelles Skimmer (4)

*Orthetrum stemmale wrightii*

Length: 43 mm/1·7 in  Wingspan: 66 mm/2·6 in

A widespread species, common in more open habitats in both plateau areas and in the hills, often occurring well away from water. Unusually for a dragonfly, appears to be more abundant during the drier southeast monsoon.

**IDENTIFICATION:** A very variable species, which could be confused with Blue Percher. However, it is larger and longer-winged and has a more powerful flight. It frequently rests with its abdomen pointing downwards and the wings angled forwards and downwards; this behaviour, together with its swollen thorax, gives the species a 'hump-backed' look when at rest. **DESCRIPTION:** Fresh adults **(4a)** are generally rather dark with a blackish abdomen, thorax and head, but there are characteristic greenish-yellow paired spots on the head behind the dark metallic-blue eyes, on the thorax and along the abdomen. The spots on the thorax remain prominent (although they may become bluer with age), and are a diagnostic feature. Adults gradually obtain a sky-blue abdomen, the black giving way to blue in bands and forming a characteristically stripy pattern **(4b)**. In mature specimens the abdomen is largely blue, although often has a darker tip as in Blue Percher. When seen closely, the abdomen is not as flattened vertically as in Blue Percher and the blue of the abdomen is a little darker, often with a slightly metallic sheen. The anal appendages are white. **BEHAVIOUR:** Flies strongly, often hunts insects by returning regularly to the same perch. **BREEDING:** Breeding activity has been recorded throughout the year.

ANT-LION

## Ant-lion

*Myrmeleon obscurus*

1

Length: 60mm/2·4in Wingspan: 40mm/1·6in

A widespread species in the lowlands, less common at higher altitudes. The adults are drawn to lights at night, whilst the larval stage inhabits distinctive conical depressions in areas of sand or loose soil.

**IDENTIFICATION:** Adults resemble a dull, stocky damselfly, but are generally only seen when attracted to lights at night. They are distinguished by their rather dull brown coloration and short antennae. The larva is rarely seen but lives at the bottom of a distinctive cone-shaped depression in loose soil **(1a)**.
**DESCRIPTION:** The adult is damselfly-like, with two pairs of wings and a slender abdomen, which is pinched-in at the base. The abdomen is dull olive-brown with darker bands, while the head and thorax are darker brown. There are two short antennae on the head and the eyes are dark. The larva is very small, <10mm long, with a swollen pale-brown segmented abdomen and a tiny head bearing large jaws.
**BEHAVIOUR:** Adults flutter helplessly around lights at night. The larva inhabits loose soil at the base of a conical pit, which is used to trap small insects, especially ants. The larvae 'spit' soil at prey to knock them to the base of the pit.

BEETLES

## Coconut Beetle

*Orcytes monoceros* Makabe

2

Length: 55mm/2·2in

A widespread and distinctive beetle, which causes extensive damage to coconut palms.

**IDENTIFICATION:** A large, black, shiny beetle, the male of which is easily identified by a 'rhinoceros'-like horn protruding from the top of its head.
**DESCRIPTION:** The carapace is entirely glossy black, lacking any obvious markings. Males bear a distinct horn on the head, but this is much reduced in females. The top of the head is slightly concave. The legs bear numerous spines and have a distinct hook at the tip.
**BEHAVIOUR:** The large, pale, dark-faced grubs burrow through the softer parts of young coconuts causing extensive damage. Adult beetles also frequent coconut palms. Adults are drawn to lights at night, and then crash about noisily, flying in ever-decreasing circles around the light.

## E Click Beetle +

*Alaus scotti*

3

Length: 30mm/1·2in

A widespread species, which is drawn to lights at night, and emits a characteristic clicking sound.

**IDENTIFICATION:** A moderately large beetle, easily identified by its largely grey-brown colour, large head shield bearing backward-pointing spikes, and curious 'flicking and clicking' behaviour when threatened.
**DESCRIPTION:** A mid-sized beetle, with a slightly shiny grey-brown carapace, which is streaked paler above. The underside is blacker. The head shield is very large, with a blunt front edge and a pronounced backward-pointing spike on each of the two rear corners. The black eyes are positioned well forward, between two relatively short (*ca.* 10mm) segmented antennae.
**BEHAVIOUR:** Often drawn to lights at night, and then blunders about helplessly. It emits distinctive clicks by flicking its body when threatened; this can cause it to leap erratically, especially when attempting to turn itself off its back. It also tucks its legs away neatly on its underside when threatened, giving it a very smooth profile.
**VOICE:** Emits a curious loud clicking sound, particularly when threatened, which may act as a defence mechanism.

1

1a

2

3

## COCKROACH

### I American Cockroach 1

*Periplanetta americana*

Length: 75 mm/3 in

A widespread introduced cockroach, which is often abundant in and around buildings.

**IDENTIFICATION:** The largest cockroach in Seychelles, easily separated from other species by its large size and amber-brown coloration. **DESCRIPTION:** Predominantly amber-brown in colour, with a paler head bearing long fawn antennae. The brown legs bear a set of spikes on the upper sections, whilst there is a distinctive spike projecting from either side of the tail-tip. **BEHAVIOUR:** Mostly nocturnal, hiding in crevices in walls and in cupboards during the day. Scampers across the floor and up walls at night; may be attracted to lights at night, when they fly around blindly.

## BEES & WASPS

### Carpenter Bee 2

*Xylocopa caffra* — Mous Brenzel

Length: 50 mm/2 in

A widespread and very distinctive 'bumble' bee, which favours cleared areas and woodland edges, where there are flowering plants. The creole name means 'aubergine bee'.

**IDENTIFICATION:** A large 'bumble' bee with a distinctive broad black abdomen that bears a bold yellow band across the middle of its upper surface. **DESCRIPTION:** The body is large and rounded and mostly black but with a broad yellow stripe across the upper surface towards the front of the abdomen and a second thinner yellow stripe across the rear of the thorax; these markings merge to form one broad yellow band when seen in flight. Some individuals show a much reduced or even absent yellow band. The wings are small and dark and show an iridescent blue sheen in direct light. The front parts of bees are often covered in pollen giving them a pale-headed appearance. **BEHAVIOUR:** A solitary bee, which flies strongly with a loud buzzing of wings and frequents flowering plants.

### Yellow Wasp 3

*Polystes olivaceus* — Mous Zonn

Length: 40 mm/1·6 in

A widespread and often numerous wasp that particularly favours more open areas.

**IDENTIFICATION:** A relatively large colonial wasp, with an unpleasant sting, which is readily identified by its typical wasp-like appearance with tapering abdomen, generally yellow colour and long yellow legs that dangle in flight. **DESCRIPTION:** Almost entirely yellow with yellow brown stripes across abdomen, and long dangling yellow legs. **BEHAVIOUR:** Adults forage actively in low vegetation, flying low and stopping to feed on flowering plants. Colonies produce a typical multi-celled 'paper' nest, which is usually placed in the open branches of a tree.

### E Potter Wasp 4

*Delta alluaudi* — Mous Masonn

Length: 30 mm/1·2 in

A widespread and distinctive wasp that builds characteristic mud nests in rock crevices and inside houses.

**IDENTIFICATION:** A slim, mid-sized wasp, with an orange-brown head and thorax, and a globular black abdomen with a very constricted base and a pointed tip, best identified by its habit of building rounded mud chambers for its grubs. **DESCRIPTION:** The head and thorax are orange-brown and the abdomen glossy blackish in colour. There is a very narrow constriction between the thorax and the bulbous, point-ended abdomen. The eyes are blackish, and the dark wings shine with an iridescent blue sheen in direct sunlight. **BEHAVIOUR:** Typically builds rounded mud chambers for its grubs on walls or even inside houses, and often on top of existing chambers. Each chamber is filled with moth larvae upon which the single wasp grub feeds.

1

2

4

3

## MILLIPEDE, CENTIPEDE & WHIP SCORPION

### E Giant Millipede 1

*Sechelleptus seychellarum* Gro Milpat

Length: 280 mm/11 in

A large millipede of wooded habitats, largely restricted to smaller offshore seabird islands where it reaches impressive densities, but also occurs on Silhouette. It emits a strongly alkaline orange-brown solution when alarmed, which can burn the skin.

**IDENTIFICATION:** Readily identified by its very large size, rounded shape in cross-section, and shiny blackish-brown coloration with many warm-brown legs.
**DESCRIPTION:** A very large millipede, with a shiny, dark-brown, segmented body, each segment becoming darker and blackish towards the rear. The numerous legs project downwards beneath the body and are rusty-brown. There are short, curved, jointed antennae on the head.
**BEHAVIOUR:** Hides in shaded areas during the heat of the day, often congregating in large numbers on the shaded trunks of trees, or in rock crevices. It is active during the cooler parts of the day, and more especially at night, when it descends to the ground and trundles around in search of food. The young live deep in the soil and leaf-litter during the day, emerging at night to feed. This is an important detritivore, leaving behind numerous rounded droppings.

### Giant Centipede 2

*Scolopendra subspinipes* Sanpye

Length: 200 mm/8 in

A very large and fearsome centipede capable of inflicting a nasty bite, that is often common close to habitation. Bites are best treated with an immediate application of an acid such as vinegar or lemon juice.

**IDENTIFICATION:** Readily identified by its very large size, flattened shape in cross-section, shiny, warm-brown body with paler legs, rapid twisting gait and large jaws.
**DESCRIPTION:** A very large centipede, with a glossy, warm-brown, segmented body. The segments are large and the body rather flattened. The numerous legs project sideways from the body and are paler brown. The head bears mid-length, straight antennae and large curved jaws.
**BEHAVIOUR:** Exclusively terrestrial, moving rapidly across the ground, often twisting the body from side to side like a snake, but active only at night, hiding under stones, or in wood piles during the day. It hunts insects, small reptiles, amphibians and birds, using its large jaws which are equipped to inject a strong alkaline solution to stun or kill prey. It may enter houses, particularly during wet weather.

### Whip Scorpion 3

*Tarantula (Phrynictus) scaber*

Tarantine Arme

Length: 60 mm/2·4 in Front legs: 120 mm across

A very large, fearsome-looking but harmless arachnid that occurs only on rat-free seabird islands and Silhouette.

**IDENTIFICATION:** Unlikely to be confused with anything else if seen well. The very large, bent front limbs bearing black claws are highly distinctive, as are the very long whip-like antennae, and the short rounded abdomen.
**DESCRIPTION:** A peculiar and fearsome-looking invertebrate. Largely grey-brown in colour, with very long whip-like antennae, greatly enlarged front limbs which are held bent in front of the body and bear shiny black pointed claws. The body is flattened, and the short, segmented abdomen is rounded. There is no sting as in a typical scorpion.
**BEHAVIOUR:** Exclusively nocturnal, hiding deep in rock crevices during the day. It emerges at night to feed, when it is often seen on rocks, in wells or on walls. It feeds largely on crickets, which it feels with its very long whip-like antennae. Although Whip Scorpions remain still for long periods, they are capable of moving at considerable speed when alarmed, running sideways in a crab-like fashion.

1

2

3

SPIDERS

## Palm Spider

1

*Nephila inaurata*

Bib

Length: 120mm/4·7in (including legs)

A large, long-legged spider that builds conspicuous flattened webs high in trees and across telegraph wires. It is of widespread occurrence and is often very common in lightly wooded habitats.

**IDENTIFICATION:** Females are distinctive, on account of their large size, long amber-and-black striped legs, long, slim and slightly upturned abdomen, and habit of living in large conspicuous webs.

**DESCRIPTION:** Females are large and conspicuous **(1f)**, whereas the tiny blackish males **(1m)** are cryptic and often lurk nearby in the same web. Adult females have a largely blackish-brown head and body, whilst younger females show thin patches of yellow on both surfaces of the abdomen and four white spots towards the front end of the upper surface of the abdomen. The long legs are warm amber in colour, but are blackish at the joints, giving them a striped appearance. The lower leg joints bear numerous short hairs.

**BEHAVIOUR:** Mostly diurnal, catching insects in large flattened webs spun across gaps in trees or between roadside wires. Young spiders live communally in more complex webs, and the large webs of adult females are often host to a range of other different spiders.

## Giant Crab Spider

2

*Heteropoda venatoria*

Length: 130mm/5·1in (including legs)

A very large, long-legged spider that often enters houses at night.

**IDENTIFICATION:** The male is a large brown spider with long slender legs, a broad head and short body, that hunts at night without the use of webs. They walk rapidly across the ground and up walls, but also freeze, motionless, for long periods. The female is squat, fat and hairy and lurks in crevices in rocks, walls and trees.

**DESCRIPTION:** Males are mid- to dark-brown in colour with a paler head and paler bases to the long, dark legs. Females are rich brown in colour, with thick, hairy legs.

**BEHAVIOUR:** Mostly nocturnal, hiding in crevices during the day and emerging to hunt at night. Males often occur in buildings where they forage on the floor and walls at night. Females are rarely seen and have a powerful bite.

1m
1f

2m

CRABS

## Giant Mangrove Crab + 1

*Cardisoma carnifex* Changomar

Carapace width: 20 cm/8 in

Locally abundant, inhabiting large burrows in and around mangrove swamps of larger islands.

**IDENTIFICATION:** A huge, fearsome-looking terrestrial crab that is easily identified by its large size, chunky carapace and massive pale-coloured claws with elongated pincers.

**DESCRIPTION:** A very large terrestrial crab that is cherry-brown above and paler below, with darker brown legs bearing orange patches at the joints. The out-sized claws are a pale cream colour and the eyes are stalked and dark.

**BEHAVIOUR:** This is a largely nocturnal species, although it may be seen at burrow entrances during the day. It leaves its burrow at night to graze on vegetation, which it plucks delicately with its huge claws. It may be encountered at night crossing coastal roads.

## Robber (Coconut) Crab 2

*Birgus latro*

Sipay

Length: 60 cm/24 in

Very rare in the granitic Seychelles, where it had not been recorded until the late 1990s when isolated specimens were encountered on Aride and Cousine islands. Previously abundant throughout the Aldabra group of islands, and remains so at Aldabra atoll.

**IDENTIFICATION:** The largest terrestrial crab; mature adults can be identified by size alone, measuring up to 1 m (39 in) from leg-tip to leg-tip and weighing up to 4 kg. Smaller crabs resemble larger hermit crabs, but usually do not carry a shell, have distinctly banded legs, large pincers of roughly equal size, a much broader, straighter body, and a pronounced point on the forehead.

**DESCRIPTION:** A very large crab, mostly purplish-brown in colour but with pale pincers and numerous pale bands particularly on the upper sections of the legs. The front pincers are large and of roughly equal size, whilst the second pair of legs is the longest. There is a pronounced point on the forehead, above and between the blood-red eyes. The body is broad and the rather straight abdomen is tucked under the body.

**BEHAVIOUR:** A largely nocturnal species, hiding up in a grass-lined burrow under logs or rocks during the day. In Seychelles, Robber Crabs avoid congregations of other crabs, though elsewhere in the species' range they are attracted to food in large groups. It forages mostly on the ground, and reputedly climbs trees.

**BREEDING:** Pairs mate on the beach and the female then lays her eggs in the sea. The larvae are only able to travel a relatively short distance during their 30-day pelagic phase before emerging onto land.

1

2

## Purple Hermit Crab + ①

*Coenobita brevimanus* Solda

Size: shell to 80mm//3·1 in across opening

Common on smaller offshore islands, but also occurs on the larger islands, inhabiting beach-crest vegetation and coastal woodland.

**IDENTIFICATION:** A large, terrestrial hermit crab, separated from other species by its large size with very powerful pincers and dull purple coloration.

**DESCRIPTION:** A large shell-bearing terrestrial hermit crab, largely dull purple in colour and with large pincers, the left one usually being larger than the right. The antennae are very long and the eyes are blackish.

**BEHAVIOUR:** Largely nocturnal, hiding in piles of rotting coconuts or under logs during the day, often communally. It is exclusively terrestrial, foraging on the ground. Large numbers may be attracted to a food source such as opened coconuts. It typically uses a turban shell for protection, although some large individuals use old coconuts.

## Orange Hermit Crab + ②

*Coenobita perlatus* Solda

Size: shell to 60mm/2·4in across opening

Fairly common on sandy beaches and in adjacent beach-crest vegetation on smaller offshore islands, as well as on the larger islands, but less numerous and more local than the Purple Hermit Crab.

**IDENTIFICATION:** A mid-sized, terrestrial hermit crab, separated from other species by its generally whitish coloration with orange tingeing. Adults are smaller than the Purple Hermit Crab and the pincers are less powerful.

**DESCRIPTION:** A mid-sized shell-bearing terrestrial hermit crab, mostly white in colour, but with heavy orange tingeing on the body, pincers and legs, particularly on the upper sections. There are distinctive round, white, knobbly spots on the legs. The pincers are carried well under the shell when walking and are relatively small. The long antennae are orange in colour, and the eyes are black.

**BEHAVIOUR:** Largely nocturnal, hiding in beach-crest undergrowth or under logs during the day. Larger individuals can be active during the late afternoon, when they may walk long distances in search of food. This species forages exclusively on the ground, mostly on the beach and normally uses turban shells for protection. It is generally solitary, but larger numbers gather on certain nights (higher tides with a full moon), possibly to mate.

## Pale Hermit Crab + ③

*Coenobita rugosus* Solda

Size: shell to 40mm/1·6in across opening

Fairly common above the high tide mark on sandy beaches and in adjacent beach-crest vegetation on smaller offshore islands; less common on larger islands.

**IDENTIFICATION:** A small, terrestrial hermit crab, separated from other species by its generally pale whitish coloration with greyer tingeing on the legs. Adults are markedly smaller than both the Purple and Orange Hermit Crabs.

**DESCRIPTION:** A small shell-bearing terrestrial hermit crab, mostly white in colour, but with some grey tingeing on the pincers and legs, particularly on the upper sections. The pincers are carried well under the shell when walking and are small. The short, bent antennae are whitish in colour, the eyes are black and the legs are relatively hairy.

**BEHAVIOUR:** Active day and night, but less so during the heat of the day when it hides in beach-crest undergrowth and often congregates in large numbers in damp hollows. It forages exclusively on the ground, mostly above the high-water mark on sandy beaches and in beach-crest vegetation.

1

2

3

## Horned Ghost Crab + 1

*Ocypode ceratopthalmus* Loulou Grangalo

Size: carapace width 40 mm/1·6 in

A widespread and often abundant resident of sandy beaches where they burrow conspicuously near the top of the beach.

**IDENTIFICATION:** An unmistakable crab of sandy beaches, generally pale green in colour with highly distinctive elongated straight 'horns' above each eye. They are active burrowers, producing mounds of sand from their holes towards the top of the beach.
**DESCRIPTION:** Generally pale green in colour, with yellow tinges at the joints, although the colouration can vary from darker green to purplish; young crabs are almost colourless. The straight, elongated, pale orange horns above the very large stalked eyes are diagnostic. The eyes are pale with darker centres. One claw is always larger than the other.
**BEHAVIOUR:** Active day and night, scavenging along the upper parts of sandy beaches and adjacent beach-crest. The males dig deep copulation burrows on a lunar cycle, with most digging occurring at new moon. Individuals are capable of running at speed to avoid predators, and can also dig themselves rapidly into the sand.

## Chunky Ghost Crab + 2

*Ocypode cordimana* Loulou Grangalo

Size: carapace width 40 mm/1·6 in

A widespread and often common resident of sandy beaches, which burrows conspicuously near the top of the beach.

**IDENTIFICATION:** Easily separated from the similar-sized Horned Ghost Crab by its more robust appearance, lack of horns above the black eyes and generally pale purplish-brown coloration. They burrow actively, producing mounds of sand from their holes towards the top of the beach.
**DESCRIPTION:** Generally pale purplish-brown in colour, with paler blue-grey claws, though odd individuals can be grey-green in colour. The bold, beady black eyes are on stalks but lack horns above. The carapace is rather rounded and deep-bodied, giving a distinctly thickset appearance. The claws are large and chunky, and one is always larger than the other.
**BEHAVIOUR:** Active day and night, scavenging along the upper parts of sandy beaches and adjacent beach-crest. They will hold their ground and raise their claws when cornered.

## Fiddler Crab + 3

*Uca* spp. Loulou Grangalo

Size: carapace width 30 mm/1·2 in

Widespread, frequenting sheltered muddy mangrove-lined inlets and mud-banks where they forage from burrows.

**IDENTIFICATION:** A small to mid-sized crab, the males of which are easily identified by the presence of one greatly enlarged pincer, which they wave to other crabs at the entrances to their burrows.
**DESCRIPTION:** Males are very distinctive, having one greatly enlarged pincer, which is usually orange-pink in colour and paler towards the tip. The carapace in adult males is pale turquoise and the legs are a purple-pink. Females and immatures are a dull blackish colour, sometimes with red tingeing to the lower parts of the legs, and have small pincers of equal size. The eyes are on long, slim stalks.
**BEHAVIOUR:** Males habitually wave their enlarged pincer at other crabs from close to the entrance of their burrow. They are active day and night, feeding on organic matter in muddy substrates, and retreat rapidly to the safety of their burrow when threatened.

1

2

3

## Swift-footed Crab + 1

*Grapsus tenuicrustatus* Karkasay

Size: carapace width 70 mm/2·8 in

Widespread, conspicuous and often abundant, occurring on rocky shores where they scurry rapidly across rocks when approached.

**IDENTIFICATION:** A mid-sized, diurnal crab that is readily identified by its flattened body, long legs and habit of scurrying rapidly across rocks at the water's edge.

**DESCRIPTION:** A very conspicuous crab, generally blackish above with paler greenish or yellowish markings, although some individuals have blood-red claws and legs **(1r)**. The body is flattened dorso-ventrally, which enables it to cling to the rocks without being washed off by wave action. The legs are rather long and spindly, and are paler than the carapace.

**BEHAVIOUR:** A diurnal species that habitually lives on inter-tidal rocks, where it gleans algae. It can withstand considerable wave action, but scurries rapidly when approached and may leap between rocks to hide in crevices, or into the sea.

## Red Rock Crab + 2

*Geograpsus stormi*

Size: carapace width 35 mm/1·4 in

Widespread, occurring on upper stretches of rocky shores where it hides in crevices. Also inhabits burrows in more muddy inter-tidal areas.

**IDENTIFICATION:** A small to mid-sized crab that is separated from other similar species by its rather flattened body, relatively short legs and typically bright blood-red legs and a reddish or maroon carapace.

**DESCRIPTION:** Occurs in two distinct colour phases. Typical crabs have bright blood-red legs with a darker maroon carapace that bears red marks, particularly at the fringes. Others have a darker brown carapace with lighter brown legs. The legs are relatively short and bear numerous hairs. The eyes are blackish.

**BEHAVIOUR:** Typically hides in rocky crevices and burrows near the top of shores, rarely venturing far from cover and darting back into its hole when threatened. It is active both during the day and night.

## Orange Land Crab + 3

*Geograpsus crinipes*

Size: carapace width 35 mm/1·4 in

Widespread, occurring inland from beaches where it hides in crevices.

**IDENTIFICATION:** A mid-sized to large terrestrial crab that is separated from similar species by its large flattened body, even-sized front legs and typically rather uniform brownish-orange coloration.

**DESCRIPTION:** A large terrestrial crab that is uniformly brownish-orange in colour. The carapace is usually a little darker than the legs and bears indistinct paler bars towards the sides. The legs are rather long, and the front pair bears a distinct comb-fronted plate on the upper section. The eyes are pale with dark centres.

**BEHAVIOUR:** Typically hides in rocky crevices behind the beach-crest, rarely venturing far from cover. It is active day and night, hunting smaller crabs.

1r

1

2

3

SLUGS & SNAILS

## E White Slug

1

*Vaginula seychellensis*

Length: 70 mm/2·8 in

A large and distinctive slug. On Praslin, an all-white form predominates which is conspicuous in the Coco-de-Mer Forests. Greyer forms predominate in the hill forests of Mahé and Silhouette.

**IDENTIFICATION:** A large, all-white or greyish coloured slug. The white form on Praslin is unmistakable, though the greyer forms on Mahé and Silhouette are similar in colour to the Grey Slug but have much broader, shorter bodies with a more distinctly rounded rear end. **DESCRIPTION:** This large, broad slug is either entirely creamy-white (Praslin) or greyish-brown with darker eye-stalks (Mahé and Silhouette). It has a very broad body with a round end and two large protruding eye-stalks. **BEHAVIOUR:** On Praslin, the white form seems largely confined to Vallée de Mai and adjacent areas of suitable habitat, where it is easily seen on the flowers, stems and fronds of the endemic Coco-de-Mer palm (particulaly male flowers). It is also widespread and relatively common in hill forests on Mahé and Silhouette, occurring on moist rocks and tree trunks. It is active throughout the day, although avoids sunlit areas.

## E Grey Slug +

2

*Vaginula voeltzkowi*

Length: 130 mm/5·1 in

A very large slug of variable colour that is common in hill forests on Mahé and Silhouette.

**IDENTIFICATION:** A very large slug, which frequents damp rocks and tree trunks in hill forest. The long greyish or brownish body is distinctive. **DESCRIPTION:** Rather variable in colour, ranging from fawn to dark grey, with diffuse darker mottling, sometimes appearing vaguely striped. It has a very long, relatively slim body, with a broad fringe that often appears pale. The eyes are on long, dark grey stalks with pale tips. **BEHAVIOUR:** Active during the day, especially when wet, climbing over rocks and tree-trunks. It contracts its body and retracts its eye-stalks when threatened.

## I African Land Snail

3

*Achatina immaculata* Gro Kourpa

Size: 100 × 650 mm/4 × 26 in

A very large terrestrial snail, which was introduced from Africa, and is now commonly seen in all habitats on most of the granitic islands, particularly at higher elevations.

**IDENTIFICATION:** A very large snail with a conical brown and white striped shell, which bears a very broad body whorl.
**DESCRIPTION:** The shell is striped with uneven wavy bands of brown, grey and white, and is conical bearing seven convex whorls, the body whorl being very broad. The mouth of the shell has a broad pink or purple lip, and there are strong uneven growth lines. The body is a pale brownish-yellow, which is darker on the head. **BEHAVIOUR:** Frequents moist piles of logs and leaf-litter during the heat of the day, coming out to feed when it is overcast or cool.

## E Coco-de-Mer Snail +

4

*Stylodonta studeriana* Kourpa

Size: 45 × 60 mm/1·8 × 2·4 in

A large and distinctive snail that is confined to mature Coco-de-Mer forest in the southern half of Praslin.

**IDENTIFICATION:** A large brown snail with a rather flattened and rounded shell.
**DESCRIPTION:** The shell is yellowish to orange-brown, and is rounded and flattened bearing six convex whorls. The mouth of the shell has a broad white lip, and there are ribs on the first three whorls. The body is brown and matches the colour of the shell. **BEHAVIOUR:** Occurs only in the Vallée de Mai and adjacent areas of suitable habitat, where it is restricted to the stems and fronds of the endemic Coco-de-Mer palm. It is active during the day and is easily seen within its extremely limited range.

1
2
3
4

# Glossary

| | |
|---|---|
| **Abdomen** | The bottom or lower part of an animal's body. |
| **Anal appendages** | The claspers at the end of the abdomen of male dragonflies with which they grasp the female during mating. |
| **Arboreal** | Living in trees. |
| **Blow hole** | The nostril of a dolphin or whale, situated on top of its head. Expulsion of air and water through this by the larger whales creates a characteristic fountain or 'blow', visible from some distance. |
| **Carapace** | The tough shell of turtles, terrapins, tortoises and some invertebrates, such as many crabs and beetles. |
| **Chevron** | Crescent or V- shaped patch of colour. |
| **Colonial** | Breeding in groups or 'colonies'. |
| **Crepuscular** | Mostly active at dawn and dusk. |
| **Detritivore** | An animal that feeds on rotting plant and animal remains. |
| **Diagnostic** | A feature that is unique to a given species. |
| **Diurnal** | Active during the day. |
| **Endemic** | Occurs only within a restricted region. |
| **Introduced** | Species whose natural range lies elsewhere but has been brought to an area by man. |
| **Leguminous** | Plants of the Legume family such as peas, brooms and vetches. |
| **Mantle** | The back of a bird or mammal. |
| **Migrant** | A species that undertakes periodic movements to and from an area at predictable times of year, normally along well-defined routes. |
| **Nocturnal** | Active at night. |
| **Offshore** | Deeper waters away from the coast. |
| **Plastron** | The hard plate on the underside of a turtle, terrapin or tortoise. |
| **Plateau** | A flat, low-lying area close to the coast. |
| **Pod** | A co-ordinated group of whales or dolphins. |
| **Raptor** | A bird of prey. |
| **Resident** | Present throughout the year at a given location. |
| **Roost** | A gathering of a species at a location to rest (either by day or at night). |
| **Rump** | The area above the tail on a bird or mammal |
| **Subspecies** | A population of a species which differs in size, appearance or call from other populations of the same species (also known as 'race'). |
| **Tadpole** | The larval form of amphibians such as frogs. |
| **Teneral** | A newly-emerged dragonfly or damselfly, which lacks the colour of the adult and often has very shiny wings. |
| **Terrestrial** | Living on the ground. |
| **Thorax** | The part of an animal's body that lies between the head and the abdomen. |
| **Vagrant** | An individual that occurs outside the normal range of the species. |

# Further Reading

BirdLife International. 2000. *Threatened Birds of the World.* Lynx Edicions & BirdLife International.

Blackman, R. and Pinhey, E. 1967. Odonata of Seychelles. *Arnoldia* (Rhodesia) 3:1-38.

Bullock, I. 1990. *Birds of the Republic of Seychelles.* Lucas Graphics Ltd.

Gerlach, J. 1997. *Seychelles Red Data Book.* The Nature Protection Trust of Seychelles.

Gerlach, J. 1987. *The Land Snails of Seychelles.* A field guide. Weedon, UK.

Harrison, P. 1983. *Seabirds an identification guide.* Croom Helm, Beckenham, UK.

Marchant, J., Prater, T. and Hayman, P. 1986. *Shorebirds an identification guide to the waders of the world.* Croom Helm, Beckenham, UK.

Nussbaum, R. A. and Brodie, E.D. (Unpublished). *The Amphibians and Reptiles of the Seychelles Archipelago.* University of Michigan, Unpublished Report from 1985.

Penny, M. 1974. *The Birds of Seychelles and Outlying Islands.* Collins, London.

Richmond, M.D. (Ed.). 1997. *A guide to the Seashores of Eastern Africa, and the Western Indian Ocean Islands.* Sida, SAREC.

Sinclair, I. and Langrand, O. 1998. *Birds of the Indian Ocean Islands.* Struik, Cape Town.

Skerrett, A. (Ed.) 1991. *Spectrum Guide to Seychelles.*

Skerrett, A. 1996. The first report of the Seychelles Birds Records Committee. *Bulletin of the African Bird Club* 3:1 45-50.

Skerrett, A. and Bullock, I. 1992. *A Birdwatchers' guide to Seychelles.* Prion, Perry, UK.

Skerrett, A., Bullock, I. and Disley, T. 2001. *Birds of Seychelles.* Helm Field Guides, A&C Black, London.

Stoddart, D.R. (Ed.). 1984. *Biogeography and Ecology of the Seychelles Islands.* Junk, The Hague.

Also volumes of the NPTS journals *Phelsuma* and *Birdwatch.*

## Useful Addresses

| | |
|---|---|
| **Aride Island Nature Reserve**<br>via GPO Grand Anse, Praslin, Seychelles. | **www.arideisland.net** |
| **Island Conservation Society Seychelles**<br>P.O. Box 775, Victoria, Seychelles. | **www.islandconservation.net** |
| **Marine Conservation Society**<br>P.O. Box 1299, Victoria, Seychelles. | **www.mcss.sc** |
| **Marine Parks Authority**<br>P.O. Box 1240, Victoria, Seychelles. | |
| **Ministry of Environment**<br>P.O. Box 445, Victoria, Seychelles. | **www.pps.gov.sc/enviro** |
| **Nature Protection Trust of Seychelles**<br>P.O. Box 207, Victoria, Seychelles. | **www.members.aol.com/jstgerlach** |
| **Nature Seychelles**<br>P.O. Box 207, Victoria, Seychelles. | **nature@seychelles.net** |
| **Plant Conservation Action Group**<br>P.O. Box 392, Victoria, Seychelles. | |
| **Seychelles Birds Records Committee**<br>P.O. Box 336, Victoria, Seychelles. | **www.stokecoll.ac.uk/sbrc** |
| **Seychelles Islands Foundation**<br>P.O. Box 853, Victoria, Seychelles. | **www.sif.sc** |
| **Wildlife Clubs of Seychelles**<br>c/o Nature Seychelles, P.O. Box 1310, Victoria, Seychelles. | **nature@seychelles.net** |

# Checklist and distribution of regularly occurring species

KEY: **Ma** = Mahé **Pr** = Praslin **Si** = Silhouette **LD** = La Digue

**SMALLER ISLANDS***

*Islands = Aride, Bird, Cerf, Conception, Cousin, Cousine, Curieuse, Denis, Félicité, Frégate, Grand Soeur, Marianne, Moyenne, North, Recif, St. Anne, Thérèse

Islets = Booby, Ile Aux Vaches Marines (off Mahé), Chauve-Souris (off Praslin), L'Ilot Frégate, Mamelles, Petite Soeur, Zave

■ = Endemic species ● = Endemic subspecies ● = Introduced species

**Y** = Occurs n = Does not occur **ex** = extinct **m** = migrant ? = Status uncertain

R = Reintroduced c = Captive

| | | Ma | Pr | Si | LD | SMALLER ISLANDS* |
|---|---|---|---|---|---|---|
| **BIRDS** | | | | | | |
| Wedge-tailed Shearwater | *Puffinus pacificus* | y? | | | | Offshore all islands, Breeds Aride, Cousin, Cousine, Bird, Recif |
| Audubon's Shearwater | *Puffinus lherminieri* | y? | | | | Offshore all islands, Breeds Aride, Cousin, Cousine, Recif |
| Red-tailed Tropicbird | *Phaethon rubricauda* | | | | | Aride only |
| White-tailed Tropicbird | *Phaethon lepturus* | Y | Y | Y | Y | Breeds Aride, Cousin, Cousine, Bird, Denis, Frégate, Mamelles, Recif |
| Great Frigatebird | *Fregata minor* | m | m | m | m | Migrant to all islands |
| Lesser Frigatebird | *Fregata ariel* | | | | | Scarce migrant to all islands, mostly Aride, Bird |
| Red-footed Booby | *Sula sula* | | | | | Rare migrant to coasts esp Aride, Bird |
| Masked Booby | *Sula dactylatra* | | | | | Rare migrant to coasts |
| Brown Booby | *Sula leucogaster* | | | | | Rare migrant to coasts esp Aride |
| Yellow Bittern | *Ixobrychus sinensis* | Y | Y | n | Y | Migrant to Aride, Curieuse |
| ● Green-backed Heron | *Butoroides striatus degens* | Y | Y | Y | Y | All but Bird+Denis |
| Black-crowned Night-Heron | *Nycticorax nycticorax* | Y | m | Y | n | Also Cousin, vagrant elsewhere |
| ● Cattle Egret | *Bubulcus ibis seychellarum* | Y | Y | n | n | Also Cousin, vagrant elsewhere |
| Grey Heron | *Ardea cinerea* | Y | m | m | n | Vagrant elsewhere |
| Purple Heron | *Ardea purpurea* | | | | | Vagrant, including Bird |
| Moorhen | *Gallinula chloropus* | Y | Y | Y | Y | Aride, Cousin, Cousine, Curieuse, Frégate, Gd Soeur, North |
| Garganey | *Anas querquedula* | m | m | m | m | Scarce migrant to all |
| Common Snipe | *Gallinago gallinago* | | | | | Rare migrant to all |
| Bar-tailed Godwit | *Limosa lapponica* | m | m | m | m | Scarce migrant to all |
| Whimbrel | *Numenius phaeopus* | m | m | m | m | Common migrant to all |
| Eurasian Curlew | *Numenius arquata* | m | m | m | m | Scarce migrant to all |
| Greenshank | *Tringa nebularia* | m | m | m | m | Migrant to all (small numbers) |
| Ringed Plover | *Charadrius hiaticula* | m | m | m | m | Scarce migrant to all |
| Greater Sandplover | *Charadrius leschenaultii* | m | m | m | m | Migrant to all |
| Lesser Sandplover | *Charadrius mongolus* | m | m | m | m | Scarce migrant to all |
| Pacific Golden Plover | *Pluvialis fulva* | m | m | m | m | Scarce migrant to all |
| Grey Plover | *Pluvialis squatarola* | m | m | m | m | Common migrant to all |
| Turnstone | *Arenaria interpres* | m | m | m | m | Common migrant to all |
| Crab Plover | *Dromas ardeola* | m | m | m | m | Migrant to all |
| Sanderling | *Calidris alba* | m | m | m | m | Common migrant to all |

| | | | Ma | Pr | Si | LD | SMALLER ISLANDS* |
|---|---|---|---|---|---|---|---|
| | Little Stint | *Calidris minuta* | m | m | m | m | Scarce migrant to all |
| | Curlew Sandpiper | *Calidris ferruginea* | m | m | m | m | Migrant to all |
| | Wood Sandpiper | *Tringa glareola* | m | m | m | m | Scarce migrant to all |
| | Terek Sandpiper | *Xenus cinereus* | m | m | m | m | Migrant to all (small numbers) |
| | Common Sandpiper | *Actitis hypoleucos* | m | m | m | m | Common migrant to all |
| | Lesser Black-backed (Baltic) Gull | *Larus fuscus* | | | | | Rare migrant to all |
| | Antarctic Skua | *Catharacta antarctica* | | | | | Scarce migrant to coastal waters |
| | Gull-billed Tern | *Geochelidon nilotica* | m | m | m | m | Scarce migrant to all |
| | Crested Tern | *Sterna bergii* | m | m | m | m | Migrant to all |
| | Lesser Crested Tern | *Sterna bengalensis* | m | m | m | m | Scarce migrant to coasts |
| | Black-naped Tern | *Sterna sumatrana* | | | | | Very scarce migrant to coasts |
| | Common Tern | *Sterna hirundo* | m | m | m | m | Scarce migrant to all coasts |
| | Roseate Tern | *Sterna dougallii* | | | | | Breeds Aride and Booby only, vagrant Bird |
| | Saunders' Tern | *Sternula saundersi* | m | m | m | m | Scarce migrant to all coasts |
| | White-winged Tern | *Chlidonias leucopterus* | m | m | m | m | Scarce migrant to all |
| | Fairy Tern | *Gygis alba* | Y | m | m | Y | Also Aride, Bird, Cousin, Cousine, Frégate |
| | Bridled Tern | *Onychoprion anaethetus* | m | m | m | m | Migrant all, Breeds Aride, Booby, Cousin, Cousine, Recif |
| | Sooty Tern | *Onychoprion fuscata* | m | m | m | m | Migrant all, Breeds Aride, Bird, Booby, Recif, L'Ilot Frégate |
| | Lesser Noddy | *Anous tenuirostris* | m | m | m | m | Migrant all, breeds Aride, Booby, Cousin, Cousine, Recif, Bird, Frégate |
| | Brown Noddy | *Anous stolidus* | | | | | Migrant all, breeds Aride, Booby, Cousin, Cousine, Recif, Bird, Frégate, L'Ilot Frégate, Vaches, Zave |
| ■ | Seychelles Kestrel | *Falco araea* | Y | Y* | Y | ex | *Rare. Also North |
| | Cuckoo | *Cuculus canorus* | | | | | Vagrant to all |
| | Roller | *Coracias garrulus* | | | | | Vagrant to all |
| ● | Seychelles Black Parrot | *Coracopsis [nigra] barklyi* | n | Y | n | n | Also Curieuse (rare), extinct Aride and Marianne |
| ■ | Seychelles Blue Pigeon | *Alectroenas pulcherrima* | Y | Y | Y | Y | All except Bird, Booby, Recif |
| ?● | Madagascar Turtle Dove | *Streptopelia picturata picturata* | Y | Y | Y | Y | All except Recif, Booby |
| ● | Seychelles Turtle Dove | *Streptopelia picturata rostrata* | n | n | n | n | Only Aride, Bird, Cousin, Cousine |
| ● | Barred Ground Dove | *Geopelia striata* | Y | Y | Y | Y | All except Recif, Booby |
| ● | Barn Owl | *Tyto alba* | Y | Y | Y | Y? | Also Aride, Curieuse, North |
| ■ | Seychelles Scops-owl | *Otus insularis* | Y | n | n | n | No |
| ■ | Seychelles Swiftlet | *Aerodramus elaphrus* | Y | Y | n | Y | Vagrant Aride, extinct Félicité |
| | Barn Swallow | *Hirundo rustica* | | | | | Vagrant to all |
| | Tree Pipit | *Anthus trivialis* | | | | | Vagrant to all |
| ■ | Seychelles Paradise Flycatcher | *Terpsiphone corvina* | n | ex | n | Y | Also Marianne, vagrant Félicité |
| ■ | Seychelles Bulbul | *Hypsipetes crassirostris* | Y | Y | Y | Y | No |
| ■ | Seychelles Magpie-robin | *Copsychus sechellarum* | n | n | n | n | Frégate, reintroduced Aride, Cousin, Cousine |
| ■ | Seychelles Warbler | *Acrocephalus sechellensis* | n | n | n | n | Cousin, reintroduced Aride, Cousine, introduced Denis |
| ■ | Seychelles Fody | *Foudia sechellarum* | n | n | n | n | Frégate, Cousin, Cousine., introduced Aride, Denis |
| ● | Madagascar Fody | *Foudia madagascariensis* | Y | Y | Y | Y | All |
| ■ | Seychelles Sunbird | *Nectarinia dussumieri* | Y | Y | Y | Y | All except Recif, Bird, Denis |
| ■ | Seychelles White-eye | *Zosterops modesta* | Y* | n | n | n | *Rare. Also Conception, introduced Frégate |
| ● | Common Myna | *Acridotheres tristis* | Y | Y | Y | Y | All except Aride, Recif |
| ● | Common Waxbill | *Estrilda astrild* | Y | n | n | Y | No |
| ● | Feral Pigeon | *Columba livia* | Y | Y | Y | Y | Vagrant to all |

| | | Ma | Pr | Si | LD | SMALLER ISLANDS* |
|---|---|---|---|---|---|---|
| **MAMMALS** | | | | | | |
| ■ Seychelles Fruit Bat | *Pteropus seychellensis* | Y | Y | Y | Y | +North. Visits all but Bird, Denis and Recif |
| ■ Sheath-tailed Bat | *Coleura seychellensis* | ex? | ex? | Y | ex? | |
| ● Ship Rat | *Rattus rattus* | Y | Y | Y | Y | All but Aride, Bird, Cousin, Cousine, Curieuse, Denis, Frégate |
| ● Norway Rat | *Rattus norvegicus* | Y | y? | n | y? | Conception (off Mahé) only, extirpated Frégate |
| ● House Mouse | *Mus domesticus* | Y | Y | Y | Y | All but Cousin, Cousine, Recif |
| ● Tenrec | *Tenrec ecaudatus* | Y | Y | n | n | None |
| ● European Rabbit | *Oryctolagus cuniculus* | y? | n | n | n | Only Recif, Mamelles and Chauve-Souris (off Praslin) |
| ● Black-naped Hare | *Lepus nigrocollis* | n | n | n | n | Only Cousin |
| **Cetaceans** | | | | | | |
| Sperm Whale | *Physeter macrocephalus* | | | | | Offshore only |
| Short-finned Pilot Whale | *Globicephala macrorynchus* | | | | | Inshore and offshore |
| Spinner Dolphin | *Stenella longirostris* | | | | | Inshore (rare) |
| Bottle-nosed Dolphin | *Tursiops truncatus* | | | | | Inshore (common) |
| Risso's Dolphin | *Grampus griseus* | | | | | Offshore only |
| **AMPHIBIANS** | | | | | | |
| **Frogs** | | | | | | |
| ■ Seychelles Tree Frog | *Tachynemis seychellensis* | Y | Y | Y | Y | Not on smaller islands (No) |
| Mascarene Frog | *Ptychadena mascariensis* | Y | Y | Y | Y | Also Curieuse, Frégate, North, Félicité, Gd Soeur, Cerf, St. Anne |
| ■ Thomasset's Frog | *Nesomantis thomasseti* | Y | n | Y | n | No |
| ■ Gardiner's Frog | *Sooglossus gardineri* | Y | n | Y | n | No |
| ■ Seychelles Frog | *Sooglossus sechellensis* | Y | n | Y | n | No |
| ■ Palm Frog | *Sooglossus pipilodryas* | n | n | Y | n | No |
| **Caecilians** | | | | | | |
| ■ Big-headed Caecilian | *Grandisonia alternans* | Y | Y | Y | Y | Also Frégate and Félicité |
| ■ Small-headed Caecilian | *Grandisonia brevis* | Y | n | Y | n | No |
| ■ Flat-headed Caecilian | *Grandisonia larvata* | Y | Y | Y | Y | Also Félicité and St. Anne |
| ■ Praslin Flat-headed Caecilian | *Grandisonia diminutiva* | n | Y | n | n | No |
| ■ Pin-headed Caecilian | *Grandisonia sechellensis* | Y | Y | Y | n | No |
| ■ Swimming Caecilian | *Hypogeophis rostratus* | Y | Y | Y | Y | Also Cerf, Curieuse, Félicité, Frégate, Gd Soeur, St. Anne |
| ■ Toothy Caecilian | *Praslina cooperi* | Y | Y | Y | n | No |
| **REPTILES** | | | | | | |
| **Tortoises** | | | | | | |
| ● Aldabra Giant Land Tortoise | *Dipsochelys dussumieri* | c | c | c | c | Introduced Cerf, Curieuse, Cousin, Cousine, Frégate, North |
| ■ Seychelles Giant Tortoise | *Dipsochelys hololissa* | | | c | | Captive only |
| ■ Arnold's Tortoise | *Dipsochelys arnoldi* | | | c | | Captive only |
| **Turtles** | | | | | | |
| Hawksbill Turtle | *Eretmochelys imbricata* | Y | Y | Y | Y | Inshore all islands |
| Green Turtle | *Chelonia mydas* | | | | | Rare inshore, breeds Aride, Cousine, Bird, Denis, St. Anne |
| ■ Seychelles Mud Turtle | *Pelusios seychellensis* | ex? | n | n | n | No |
| ● Seychelles Yellow-bellied Mud Turtle | *Pelusios castanoides intergularis* | Y | Y | n | Y | Also Curieuse (extinct?), Cousin (extinct?) |
| ● Seychelles Black Mud Turtle | *Pelusios subniger parietalis* | Y | Y | R | Y | Also Frégate and Cerf |

| | | | Ma | Pr | Si | LD | SMALLER ISLANDS* |
|---|---|---|---|---|---|---|---|
| | **Geckos** | | | | | | |
| ■ | Stripeless Day Gecko | *Phelsuma astriata* | Y | Y | Y | Y | Aride, Cousin, Cousine, Félicité, Frégate, Grand Soeur, Marianne, North |
| ■ | Long Island Day Gecko | *Phelsuma (sundbergi) longinsulae* | Y | n | Y | n | Bird, North, Frégate, all small islands around Mahé |
| ■ | Sundberg's Day Gecko | *Phelsuma sundbergi* | n | Y | n | Y | Curieuse, Félicité, Marianne, Gd Soeur |
| ■ | Seychelles Sucker-tailed Gecko | *Urocotyledon inexpectata* | Y | Y | Y | Y | All but Bird, Denis |
| ● | Pacific Gecko | *Gehyra mutilata* | Y | Y | Y | Y | All but Aride and Cousin |
| ● | Common House Gecko | *Hemidactylus frenatus* | n | n | n | n | Bird only (although could appear elsewhere) |
| ■ | Bronze Gecko | *Ailuronyx seychellensis* | Y | Y | Y | Y | All but Bird, Denis, Curieuse, Gd Soeur, Marianne |
| ■ | Dwarf Bronze Gecko | *Ailuronyx tachyscopaeus* | Y | Y | Y | Y | No |
| ■ | Giant Bronze Gecko | *Ailuronyx trachygaster* | n | Y | Y | n | No |
| | **Chameleons** | | | | | | |
| ■ | Tiger Chameleon | *Chamaeleo tigris* | Y | Y | Y | n | No |
| | **Skinks** | | | | | | |
| ■ | Seychelles Skink | *Mabuya sechellensis* | Y | Y | Y | Y | All but Recif, Mamelles, Ile aux Vaches (off Mahé) |
| ■ | Wright's Skink | *Mabuya wrightii* | n | n | n | n | Aride, Cousin, Cousine, Frégate, Recif, Booby, Ile aux Vaches, Mamelles, L'Ilot Frégate |
| ■ | Brauer's Skink | *Janetaescincus braueri* | Y | n | Y | Y | Also Frégate and Félicité |
| ■ | Gardiner's Skink | *Pamelascincus gardineri* | Y | Y | Y | Y | Also Aride, Cousin, Curieuse, Grand + Petit Soeur, Félicité, Frégate |
| | **Snakes** | | | | | | |
| ■ | Seychelles House Snake | *Lamprophus geometricus* | Y | Y | Y | n | Also Frégate |
| ■ | Seychelles Wolf Snake | *Lycognathophis seychellensis* | Y | Y | Y | n | Also Aride and Cousine |
| ● | Brahminy Blind Snake | *Ramphotyphlops braminus* | Y | Y | Y | Y | All inhabited islands inc Bird, Denis, North + Recif |
| | **INSECTS** | | | | | | |
| | **Butterflies** | | | | | | |
| | African Monarch | *Danaus chrysippus* | m | m | m | m | Migrant to all |
| ■ | King of Seychelles | *Euploea mitra* | Y | n | Y | n | No |
| | Diadem | *Hypolimnas misippus* | m | m | m | m | Migrant to all |
| | Painted Lady | *Vanessa cardui* | m | m | m | m | Migrant to all |
| | Evening Brown | *Melanitis leda* | Y | Y | Y | n | Also Aride and North |
| | Striped Policeman | *Coeliades forestan* | Y | Y | n | Y | Also Aride |
| | Olive-haired Swift | *Borbo borbonica* | Y | Y | Y | n | Also Aride and Cousine |
| | Long-tailed Blue | *Lampides boeticus* | Y | Y | Y | Y | Also Aride and Cousine |
| | Lang's Short-tailed Blue | *Syntarucus pirithous* | Y | Y | n | Y | Also Aride |
| | African Grass Blue | *Zizeeria knysa* | Y | Y | Y | Y | All but Recif, Booby, Bird, Denis |
| | Tiny Grass Blue | *Zizula hylax* | Y | Y | Y | Y | Also Aride |
| | **Moths** | | | | | | |
| | Convolvulus Hawk-moth | *Agrius convolvuli* | m | m | m | m | Migrant to all |
| | Oleander Hawk-moth | *Daphnis nerii* | m | m | m | m | Migrant to all |
| | **Damselflies** | | | | | | |
| ■ | Black Mountain Damsel | *Allolestes maclachlanii* | Y | Y | Y | n | No |
| ■ | Seychelles Stream Damsel | *Leptocnemis cyanops* | Y | Y | Y | n | No |
| | Common Pond Damselfly | *Ceriagrion glabrum* | Y | Y | Y | Y | Also Aride, Cerf, Cousin, Cousine, Curieuse, Grand Soeur, St. Anne, North |
| | Marsh Bluetail | *Ischnura senegalensis* | Y | Y | n | Y | Also Aride, Curieuse, North, |
| | Pygmy Wisp | *Agriocnemis pygmaea* | Y | Y | Y | Y | Also Aride |
| | Oceanic Fineliner | *Teinobasis alluaudi* | Y | n | Y | n | No |

| | | Ma | Pr | Si | LD | SMALLER ISLANDS* |
|---|---|---|---|---|---|---|
| **Dragonflies** | | | | | | |
| Ghost Duskdarter | *Zyxomma petiolatum* | Y | Y | Y | Y | Also Aride, Cousin, St. Anne |
| ■ Seychelles Cascader | *Zygonyx luctifera* | Y | Y | Y | n | Also St. Anne |
| Phantom Flutterer | *Rhyothemis semihylina* | Y | Y | n | Y | Also Aride,Curieuse |
| Black Percher | *Diplacodes lefebvrii* | Y | Y | n | Y | Also Aride and North |
| Voyaging Glider | *Tramea limbata* | Y | Y | Y | Y | Also Aride, Curieuse, |
| Vagrant Emperor | *Anax ephippiger* | Y | m? | m? | m? | Also Aride, Frégate but migrant to all? |
| Wandering Glider | *Pantala flavescens* | Y | Y | Y | Y | Also Aride, Gd Soeur, Curieuse, North, migrant to all? |
| Twister | *Tholymis tillarga* | Y | Y | Y | Y | Also Aride, Cerf, Frégate, North |
| ■ Seychelles Duskhawker | *Gynacantha stylata* | Y | Y | Y | n | No |
| Malagasy Emerald | *Hemicordulia similis* | m? | n | n | n | No |
| Amberwing Emperor | *Anax guttatus* | Y | Y | Y | Y | Also Aride, Frégate, North, St. Anne |
| ● Seychelles Skimmer | *Orthetrum stemmale wrightii* | Y | Y | Y | Y | All (inc Bird+Denis) except Frégate, St. Anne |
| Blue Percher | *Diplacodes trivialis* | Y | Y | Y | Y | Also Aride, Gd Soeur, Curieuse, |
| **INVERTEBRATES** | | | | | | |
| **Beetles, wasps** | | | | | | |
| Ant-lion | *Myrmeleon obscurus* | Y | Y | Y | Y | Also Aride, Bird, North |
| ■ Click Beetle | *Alaus scotti* | Y | n | Y | n | Also Aride |
| Coconut Beetle | *Orcytes monoceros* | Y | Y | Y | Y | Also Aride, Frégate, Denis |
| ● American Cockroach | *Periplanetta americana* | Y | Y | Y | Y | No? (although could appear elsewhere) |
| Yellow Wasp | *Polystes olivaceus* | Y | Y | Y | Y | All islands inc Bird+Denis |
| Carpenter Bee | *Xylocopa caffra* | Y | Y | Y | Y | Also Aride, Bird |
| ■ Potter Wasp | *Delta alluaudii* | Y | Y | Y | n? | Also Aride and Denis |
| **Millipedes, Centipede** | | | | | | |
| ■ Giant Millipede | *Sechelleptus seychellarum* | n | n | Y | Y | Aride, St. Anne, Vaches, Cousin, Cousine, Félicité, Frégate |
| Giant Centipede | *Scolopendra subspinipes* | Y | Y | Y | n | Also Aride, Frégate |
| Whip Scorpion | *Tarantula scaber* | n | n | Y | n | Also Aride, Cousin, Cousine and Frégate |
| **Spiders** | | | | | | |
| Palm Spider | *Nephila inaurata* | Y | Y | Y | Y | Also Cousin, Curieuse, Frégate, North |
| Giant Crab Spider | *Heteropoda venatoria* | Y | Y | Y | Y | No |
| **CRUSTACEANS** | | | | | | |
| Robber Crab | *Birgus latro* | n? | n? | n? | n? | Only known on Aride and Cousine |
| Giant Mangrove Crab | *Cardisoma carnifex* | Y | Y | Y | Y | No |
| Purple Hermit Crab | *Coenobita brevimanus* | Y | Y | Y | Y | All islands? |
| Orange Hermit Crab | *Coenobita perlatus* | Y | Y | Y | Y | All islands? |
| Pale Hermit Crab | *Coenobita rugosus* | Y | Y | Y | Y | All islands? |
| Fiddler Crab | *Uca* spp. | Y | Y | Y | Y | No |
| Horned Ghost Crab | *Ocypode ceratopthalmus* | Y | Y | Y | Y | All except Recif |
| Chunky Ghost Crab | *Ocypode cordimana* | Y | Y | Y | Y | All except Recif |
| Swift-footed Crab | *Grapsus tenuicrustatus* | Y | Y | Y | Y | All (inc Bird+Denis) |
| Red Rock Crab | *Geograspus stormi* | Y | Y | Y | Y | All |
| Orange Land Crab | *Geograpsus crinipes* | Y | Y | Y | Y | All |
| **MOLLUSCS** | | | | | | |
| ■ White Slug | *Vaginula seychellensis* | Y | Y | Y | n | Also North |
| ■ Grey Slug | *Vaginula voeltzkowi* | Y | n | Y | n | No |
| ■ Coco de Mer Snail | *Stylodonta studeriana* | n | n | Y | n | No |
| ● African Land Snail | *Achatina immaculata* | Y | Y | n | Y | No |

# Complete checklist of the bird species recorded in northern Seychelles (up to December 2005)

Key: E = endemic species e = endemic subspecies
B = breeding species + = extinct
M = regular non-breeding migrant R = rare migrant recorded more than twenty times
V = very rare migrant recorded 5-20 times c = escape from captivity
X = extremely rare migrant recorded less than five times.

| E/e | SPECIES | SCIENTIFIC NAME | STATUS | | | | | |
|---|---|---|---|---|---|---|---|---|
| | | | B | + | M | R | V | X |
| | Southern Giant Petrel | *Macronectes giganteus* | | | | | | ● |
| | Kermadec Petrel | *Pterodroma neglecta* | | | | | | ● |
| | Jouanin's Petrel | *Bulweria fallax* | | | | | | ● |
| | Flesh-footed Shearwater | *Puffinus carneipes* | | | | | | ● |
| | Wedge-tailed Shearwater | *Puffinus pacificus* | ● | | | | | |
| | Audubon's Shearwater | *Puffinus lherminieri* | ● | | | | | |
| | Wilson's Storm-petrel | *Oceanites oceanicus* | | | | | | ● |
| | Swinhoe's Storm-petrel | *Oceanodroma monorhis* | | | | | | ● |
| | Black-necked Grebe | *Podiceps nigricollis* | | | | | | ● |
| | Red-billed Tropicbird | *Phaethon aethereus* | | | | | | ● |
| | Red-tailed Tropicbird | *Phaethon rubricauda* | ● | | | | | |
| | White-tailed Tropicbird | *Phaethon lepturus* | ● | | | | | |
| | Red-footed Booby | *Sula sula* | | | ● | | | |
| | Masked Booby | *Sula dactylatra* | | | ● | | | |
| | Brown Booby | *Sula leucogaster* | | | ● | | | |
| | Great Cormorant | *Phalacrocorax carbo* | | | | | | ● |
| | Long-tailed Cormorant | *Phalacrocorax africanus* | | | | | | ● |
| | Great Frigatebird | *Fregata minor* | | | ● | | | |
| | Lesser Frigatebird | *Fregata ariel* | | | ● | | | |
| | Eurasian Bittern | *Botaurus stellaris* | | | | | | ● |
| | Cinnamon Bittern | *Ixobrychus cinnamomeus* | | | | | | ● |
| | Yellow (Chinese) Bittern | *Ixobrychus sinensis* | ● | | | | | |
| e | Green-backed (Striated) Heron | *Butoroides striatus degens* | ● | | | | | |
| | Black-crowned Night-heron | *Nycticorax nycticorax* | ● | | | | | |
| | Squacco Heron | *Ardeola ralloides* | | | | | | ● |
| | Indian Pond-heron | *Ardeola grayii* | | | | | | ● |
| | Madagascar Pond-heron | *Ardeola idae* | | | | | | ● |
| e | Cattle Egret | *Bubulcus ibis seychellarum* | ● | | | | | |
| | Little Egret | *Egretta garzetta* | | | | | ● | |
| | Intermediate Egret | *Egretta intermedia* | | | | | | ● |
| | Great White Egret | *Egretta alba* | | | | | | ● |
| | Grey Heron | *Ardea cinerea* | ● | | | | | |
| | Purple Heron | *Ardea purpurea* | | | | ● | | |
| | White Stork | *Ciconia ciconia* | | | | | | ● |
| | Glossy Ibis | *Plegadis falcinellus* | | | | | | ● |
| | Greater Flamingo | *Phoenicopterus ruber* | | | | | | ● |
| | Ruddy Shelduck | *Tadorna ferruginea* | | | | | | ● |
| | Mallard | *Anas platyrhynchos* | | | | | | ● |
| | Northern Pintail | *Anas acuta* | | | | | | ● |
| | Northern Shoveler | *Anas clypeata* | | | | | ● | |
| | Garganey | *Anas querquedula* | | | ● | | | |
| | Ferruginous Duck | *Aythya nyroca* | | | | | | ● |
| | Osprey | *Pandion haliaetus* | | | | | | ● |
| | Black Kite | *Milvus migrans* | | | | | | ● |

| E/e | SPECIES | SCIENTIFIC NAME | STATUS | | | | | |
|---|---|---|---|---|---|---|---|---|
| | | | B | + | M | R | V | X |
| | Marsh Harrier | *Circus aeruginosus* | | | | | | ● |
| | Booted Eagle | *Hieraaetus pennatus* | | | | | | ● |
| | Western Honey-buzzard | *Pernis apivorus* | | | | | | ● |
| E | Seychelles Kestrel | *Falco araea* | ● | | | | | |
| | Lesser Kestrel | *Falco naumanni* | | | | | | ● |
| | Eleonora's Falcon | *Falco eleonorae* | | | | | ● | |
| | Sooty Falcon | *Falco concolor* | | | | | | ● |
| | Western Red-footed Falcon | *Falco vespertinus* | | | | | | ● |
| | Amur Falcon | *Falco amurensis* | | | | | ● | |
| | Eurasian Hobby | *Falco subbuteo* | | | | | ● | |
| | Saker | *Falco cherrug* | | | | | | ● |
| | Lanner | *Falco biarmicus* | | | | | | ● |
| | Peregrine Falcon | *Falco peregrinus* | | | | | | ● |
| | Common Quail | *Coturnix coturnix* | | | | | | ● |
| | Spotted Crake | *Porzana porzana* | | | | | | ● |
| | Little Crake | *Porzana parva* | | | | | | ● |
| | Corncrake | *Crex crex* | | | | | | ● |
| | Moorhen | *Gallinula chloropus* | ● | | | | | |
| | Allen's Gallinule | *Porphyrio alleni* | | | | | | ● |
| | White-breasted Waterhen | *Amaurornis phoenicurus* | | | | | | ● |
| | Crab Plover | *Dromas ardeola* | | | ● | | | |
| | Eurasian Oystercatcher | *Haematopus ostralegus* | | | | | | ● |
| | Black-winged Stilt | *Himantopus himantopus* | | | | | | ● |
| | Stone Curlew | *Burhinus oedicnemus* | | | | | | ● |
| | Collared Pratincole | *Glareola pratincola* | | | | | | ● |
| | Oriental Pratincole | *Glareola maldivarum* | | | | | | ● |
| | Black-winged Pratincole | *Glareola nordmanni* | | | | | | ● |
| | Madagascar Pratincole | *Glareola ocularis* | | | | | | ● |
| | Little Ringed Plover | *Charadrius dubius* | | | | | ● | |
| | Ringed Plover | *Charadrius hiaticula* | | | ● | | | |
| | Lesser Sandplover | *Charadrius mongolus* | | | ● | | | |
| | Greater Sandplover | *Charadrius leschenaultii* | | | ● | | | |
| | Caspian Plover | *Charadrius asiaticus* | | | | | ● | |
| | Oriental Plover | *Charadrius veredus* | | | | | | ● |
| | Pacific Golden Plover | *Pluvialis fulva* | | | ● | | | |
| | Grey Plover | *Pluvialis squatarola* | | | ● | | | |
| | Great Knot | *Calidris tenuirostris* | | | | | | ● |
| | Broad-billed Sandpiper | *Limicola falcinellus* | | | | | | ● |
| | Sanderling | *Calidris alba* | | | ● | | | |
| | Little Stint | *Calidris minuta* | | | ● | | | |
| | Temminck's Stint | *Calidris temminckii* | | | | | ● | |
| | Long-toed Stint | *Calidris subminuta* | | | | | | ● |
| | Pectoral Sandpiper | *Calidris melanotos* | | | | | | ● |
| | Sharp-tailed Sandpiper | *Calidris acuminata* | | | | | | ● |
| | Curlew Sandpiper | *Calidris ferruginea* | | | ● | | | |
| | Ruff | *Philomachus pugnax* | | | | | ● | |
| | Buff-breasted Sandpiper | *Tryngites subruficollis* | | | | | | ● |
| | Common Snipe | *Gallinago gallinago* | | | | ● | | |
| | Great Snipe | *Gallinago media* | | | | | | ● |
| | Pintail Snipe | *Gallinago stenura* | | | | | | ● |
| | Black-tailed Godwit | *Limosa limosa* | | | | | | ● |
| E | Bar-tailed Godwit | *Limosa lapponica* | | | ● | | | |
| | Little Curlew | *Numenius minutus* | | | | | | ● |
| | Whimbrel | *Numenius phaeopus* | | | ● | | | |
| | Eurasian Curlew | *Numenius arquata* | | | ● | | | |

| E/e | SPECIES | SCIENTIFIC NAME | STATUS | | | | | |
|---|---|---|---|---|---|---|---|---|
| | | | B | + | M | R | V | X |
| | Redshank | *Tringa totanus* | | | | | | ● |
| | Spotted Redshank | *Tringa erythropus* | | | | | | ● |
| | Marsh Sandpiper | *Tringa stagnatilis* | | | | | ● | |
| | Greenshank | *Tringa nebularia* | | | ● | | | |
| | Green Sandpiper | *Tringa ochropus* | | | | | ● | |
| | Wood Sandpiper | *Tringa glareola* | | | ● | | | |
| | Terek Sandpiper | *Xenus cinereus* | | | ● | | | |
| | Common Sandpiper | *Actitis hypoleucos* | | | ● | | | |
| | Grey-tailed Tattler | *Heteroscelus brevipes* | | | | | | ● |
| | Ruddy Turnstone | *Arenaria interpres* | | | ● | | | |
| | Red-necked Phalarope | *Phalaropus lobatus* | | | | | | ● |
| | Arctic Skua | *Stercorarius parasiticus* | | | | | | ● |
| | Antarctic Skua | *Catharacta antarctica* | | | ● | | | |
| | South Polar Skua | *Catharacta maccormicki* | | | | | | ● |
| | Black-headed Gull | *Larus ridibundus* | | | | | ● | |
| | Sooty Gull | *Larus hemprichii* | | | | | | ● |
| | Lesser Black-backed (Baltic) Gull | *Larus fuscus fuscus* | | | | | ● | |
| | Siberian Gull | *Larus (fuscus) heuglini* | | | | | | ● |
| | Sooty Tern | *Onychoprion fuscata* | ● | | | | | |
| | Bridled Tern | *Onychoprion anaethetus* | ● | | | | | |
| | Little Tern | *Sternula albifrons* | | | | | | ● |
| | Saunders' Tern | *Sternula saundersi* | | | ● | | | |
| | Gull-billed Tern | *Geochelidon nilotica* | | | ● | | | |
| | Caspian Tern | *Hydroprogne caspia* | | | | | | ● |
| | Whiskered Tern | *Chlidonias hybrida* | | | | | | ● |
| | White-winged Tern | *Chlidonias leucopterus* | | | ● | | | |
| | Sandwich Tern | *Sterna sandvicensis* | | | | | | ● |
| | Crested Tern | *Sterna bergii* | | | ● | | | |
| | Lesser Crested Tern | *Sterna bengalensis* | | | ● | | | |
| | Black-naped Tern | *Sterna sumatrana* | | | | | ● | |
| | Common Tern | *Sterna hirundo* | | | ● | | | |
| | Roseate Tern | *Sterna dougallii* | ● | | | | | |
| | White-cheeked Tern | *Sterna repressa* | | | | | | ● |
| | Lesser Noddy | *Anous tenuirostris* | ● | | | | | |
| | Brown Noddy | *Anous stolidus* | ● | | | | | |
| | Fairy Tern | *Gygis alba* | ● | | | | | |
| | Feral Pigeon | *Columba livia* | ● | | | | | |
| | European Turtle Dove | *Streptopelia turtur* | | | | | | ● |
| | Madagascar Turtle Dove | *Streptopelia picturata* | ● | | | | | |
| e | Seychelles Turtle Dove | *Streptopelia picturata rostrata* | ● | | | | | |
| | Barred Ground Dove | *Geopelia striata* | ● | | | | | |
| E | Seychelles Blue Pigeon | *Alectroenas pulcherrima* | ● | | | | | |
| | Grey-headed Lovebird | *Agapornis canus* | | ● | | | | |
| e | Green Parakeet | *Psittacula eupatria wardi* | | ● | | | | |
| | Ring-necked Parakeet | *Psittacula krameri* | | | | | c | |
| e | Seychelles Black Parrot | *Coracopsis (nigra) barklyi* | ● | | | | | |
| | Jacobin Cuckoo | *Oxylophus jacobinus* | | | | | ● | |
| | Great Spotted Cuckoo | *Clamator glandarius* | | | | | | ● |
| | Common Cuckoo | *Cuculus canorus* | | | | ● | | |
| | Asian Lesser Cuckoo | *Cuculus poliocephalus* | | | | | ● | |
| | Barn Owl | *Tyto alba* | ● | | | | | |
| | Eurasian Scops-owl | *Otus scops* | | | | | | ● |
| | Seychelles Scops-owl | *Otus insularis* | ● | | | | | |
| | Brown Fish-owl | *Ketupa zeylonensis* | | | | | | ● |
| | European Nightjar | *Caprimulgus europaeus* | | | | | | ● |

| E/e | SPECIES | SCIENTIFIC NAME | STATUS | | | | | |
|---|---|---|---|---|---|---|---|---|
| | | | B | + | M | R | V | X |
| | White-throated Needletail | *Hirundapus caudacutus* | | | | | | ● |
| E | Seychelles Swiftlet | *Aerodramus elaphrus* | ● | | | | | |
| | Common Swift | *Apus apus* | | | | | ● | |
| | Pacific Swift | *Apus pacificus* | | | | | ● | |
| | Little Swift | *Apus affinis* | | | | | | ● |
| | Grey-headed Kingfisher | *Halcyon leucocephala* | | | | | | ● |
| | Blue-cheeked Bee-eater | *Merops persicus* | | | | | ● | |
| | European Roller | *Coracias garrulus* | | | | | ● | |
| | Broad-billed Roller | *Eurystomus glaucurus* | | | | | ● | |
| | Greater Short-toed Lark | *Calandrella brachydactyla* | | | | | | ● |
| | Bimaculated Lark | *Melanocorypha bimaculata* | | | | | | ● |
| | Common Sand Martin | *Riparia riparia* | | | | | ● | |
| | Mascarene Martin | *Phedina borbonica* | | | | | | ● |
| | Barn Swallow | *Hirundo rustica* | | | | ● | | |
| | Common House Martin | *Delichon urbicum* | | | | | | ● |
| | Yellow Wagtail | *Motacilla flava* | | | | | ● | |
| | Citrine Wagtail | *Motacilla citreola* | | | | | | ● |
| | Grey Wagtail | *Motacilla cinerea* | | | | | | ● |
| | White Wagtail | *Motacilla alba* | | | | | ● | |
| | Tree Pipit | *Anthus trivialis* | | | | ● | | |
| | Red-throated Pipit | *Anthus cervinus* | | | | | ● | |
| E | Seychelles Bulbul | *Hypsipetes crassirostris* | ● | | | | | |
| | Common Redstart | *Phoenicurus phoenicurus* | | | | | | ● |
| | Whinchat | *Saxicola rubetra* | | | | | | ● |
| | Northern Wheatear | *Oenanthe oenanthe* | | | | | ● | |
| | Isabelline Wheatear | *Oenanthe isabellina* | | | | | | ● |
| | Pied Wheatear | *Oenanthe pleschanka* | | | | | | ● |
| | Rock Thrush | *Monticola saxatilis* | | | | | | ● |
| E | Seychelles Magpie-robin | *Copsychus sechellarum* | ● | | | | | |
| | Sedge Warbler | *Acrocephalus schoenobaenus* | | | | | | ● |
| E | Seychelles Warbler | *Acrocephalus sechellensis* | ● | | | | | |
| | Icterine Warbler | *Hippolais icterina* | | | | | | ● |
| | Willow Warbler | *Phylloscopus trochilus* | | | | | | ● |
| | Wood Warbler | *Phylloscopus sibilatrix* | | | | | | ● |
| | Blackcap | *Sylvia atricapilla* | | | | | | ● |
| | Common Whitethroat | *Sylvia communis* | | | | | | ● |
| | Spotted Flycatcher | *Muscicapa striata* | | | | | ● | |
| E | Seychelles Paradise Flycatcher | *Terpsiphone corvina* | ● | | | | | |
| E | Seychelles Sunbird | *Nectarinia dussumieri* | ● | | | | | |
| E | Seychelles White-eye | *Zosterops modesta* | ● | | | | | |
| e | Chestnut-flanked White-eye | *Zosterops mayottensis semiflavus* | | ● | | | | |
| | European Golden Oriole | *Oriolus oriolus* | | | | | | ● |
| | Red-backed Shrike | *Lanius collurio* | | | | | | ● |
| | Woodchat Shrike | *Lanius senator* | | | | | | ● |
| | Lesser Grey Shrike | *Lanius minor* | | | | | | ● |
| | Rose-coloured Starling | *Sturnus roseus* | | | | | | ● |
| | Wattled Starling | *Creatophora cinerea* | | | | | | ● |
| | Common Myna | *Acridotheres tristis* | ● | | | | | |
| | Indian House Crow | *Corvus splendens* | | | | | | ● |
| | Common Waxbill | *Estrilda astrild* | ● | | | | | |
| | House Sparrow | *Passer domesticus* | | | | | | ● |
| | Madagascar Fody | *Foudia madagascariensis* | ● | | | | | |
| E | Seychelles Fody | *Foudia sechellarum* | ● | | | | | |
| | Common Rosefinch | *Carpodacus erythrinus* | | | | | | ● |
| | Ortolan Bunting | *Emberiza hortulana* | | | | | | ● |

# Photographic Credits

Each of the photographs and illustrations used in this book is listed in this section, together with the name of the photographer or artist. Where known, information is included on the location the photograph was taken.

### INTRODUCTORY SECTIONS

FRONT COVER:
**Seychelles White-eye** *Zosterops modesta*: David & Nancy Massie.
**Stripeless Day Gecko** *Phelsuma astriata*: John Bowler.
**Fairy Tern** *Gygis alba* in flight: Alan Tate.
**King of Seychelles** *Euploea mitra* Museum specimen: Justin Gerlach.
**Hawksbill Turtle** *Eretmochelys imbricata*: John Bowler.

BACK COVER:
**Tiger Chameleon** *Chamaeleo tigris*: John Bowler.
**John Bowler**: Janet Bowler.

TITLE PAGE:
**Seychelles Paradise Flycatcher** *Terpsiphone corvina*: Nigel McCall.

KEY TO SPECIES: All images from plates except Chameleon: John Bowler. Bird (Madagascar Fody male): Bill & Chris Wain

P.6: **Beach, Aride**: Rob Still
P.9: **Congo Rouge**: Bill & Chris Wain.
P.10: **Map of Seychelles**: ILLUSTRATION: Rob Still.
P.11: **Cousin**: Rob Still.
P.13: **Climate charts**: ILLUSTRATION: Rob Still.
P.14: **Open sea**: Off Aride: John Bowler. **Rocky islet**: Booby Island: John Bowler.
P.15: **Inter-tidal areas**: Bird and Silhouette: John Bowler.
P.16: **Lowland forest**: Aride: John Bowler. **Freshwater wetland**: Mahé: John Bowler.
P.17: **Coconut plantation**: Justin Gerlach. **Mid-altitude forest**: Praslin: John Bowler.
P.18: **High altitude forest**: Justin Gerlach.
P.19: **Supplementary feeding of Seychelles Magpie-robins**: Aride: John Bowler.
P.20: **Ringing White-tailed Tropicbirds**: Aride: Janet Bowler. **Zodiac landing**: Aride: Janet Bowler.
P.23: **Vallée de Mai**: Rob Still.
P.24: **Bird Island**: David & Nancy Massie.
P.25: **Sooty Tern colony**: Aride: John Bowler.
P.26: **Pitcher Plant**: Mahé: John Bowler.
P.28: **Seychelles Blue Pigeon**: Nigel McCall.

### THE PLATES

#### BIRDS

PLATE 1
**Audubon's Shearwater** *Puffinus lherminieri*
Immature at nest, Aride: John Bowler.
Adult in flight & group on water, Galápagos: Andy & Gill Swash.
**Wedge-tailed Shearwater** *Puffinus pacificus*
On land, Aride & bird swimming: John Bowler.
In flight: Tony Palliser.

PLATE 2
**Red-tailed Tropicbird** *Phaethon rubricauda*
In flight: Aride: John Bowler.
**White-tailed Tropicbird** *Phaethon lepturus*
All photographs, Aride: John Bowler.

PLATE 3
**Red-footed Booby** *Sula sula*
White phase in flight: Kevin Carlson (Windrush).
White phase perched (with White-tailed Tropicbird), Aride: John Bowler.
**Masked Booby** *Sula dactylatra*
In flight, Galápagos: Andy & Gill Swash.
**Brown Booby** *Sula leucogaster*
In flight, Bahamas: Bruce Hallett.

PLATE 4
**Great Frigatebird** *Fregata minor*
All birds in flight, Aride: John Bowler.
Juvenile perched, Galápagos: Andy & Gill Swash
Group soaring (facing page): Aride: John Bowler.
**Lesser Frigatebird** *Fregata ariel*
All photographs, Tuamotu Is.: Roger & Liz Charlwood.

PLATE 5
**Green-backed Heron** *Butoroides striatus degens*
Adult: Justin Gerlach.
Juvenile, Aride: John Bowler.
**Yellow Bittern** *Ixobrychus sinensis*
Adult & juvenile: Gehan de Silva.
**Black-crowned Night-heron** *Nycticorax nycticorax*
Adult & juvenile: Gehan de Silva.

PLATE 6

**Cattle Egret** *Bubulcus ibis seychellarum*
Breeding & non-breeding:, Mahé: John Bowler.

**Grey Heron** *Ardea cinerea*
Adult, South Africa: Andy & Gill Swash.
Juvenile, Mahé: John Bowler.

**Purple Heron** *Ardea purpurea*
Adult: Gehan de Silva.
In flight, South Africa: Andy & Gill Swash.

PLATE 7

**Moorhen** *Gallinula chloropus*
Adult & juvenile, Aride: John Bowler.

**Garganey** *Anas querquedula*
Male: Trevor Codlin. Female: Gehan de Silva.

**Common Snipe** *Gallinago gallinago*
Adult, UK: Andy & Gill Swash.

PLATE 8

**Greenshank** *Tringa nebularia*
Non-breeding: Greg & Yvonne Dean.

**Bar-tailed Godwit** *Limosa lapponica*
Non-breeding, UK: Peter J. Wilson.

**Whimbrel** *Numenius phaeopus*
Adult, South Africa: Andy & Gill Swash.

**Eurasian Curlew** *Numenius arquata*
Adult, UK: Andy & Gill Swash.

PLATE 9

**Ringed Plover** *Charadrius hiaticula*
Breeding: Andy & Gill Swash.
Non-breeding: Peter J. Wilson.

**Lesser Sandplover** *Charadrius mongolus*
Non-breeding, Aride: John Bowler.
Breeding: Gehan de Silva.

**Greater Sandplover** *Charadrius leschenaultii*
Non-breeding, Aride: John Bowler.
In flight: Gehan de Silva.

PLATE 10

**Grey Plover** *Pluvialis squatarola*
Moulting adult: Gehan de Silva.
Non-breeding, UK: Trevor Codlin.
Non-breeding showing axillaries: Gehan de Silva.

**Pacific Golden Plover** *Pluvialis fulva*
Breeding: Gehan de Silva.
Non-breeding, Bird: John Bowler.

**Turnstone** *Arenaria interpres*
Non-breeding, Aride: John Bowler.

**Crab Plover** *Dromas ardeola*
Adult, Aride: John Bowler.

PLATE 11

**Sanderling** *Calidris alba*
Non-breeding, running: Gehan de Silva.

**Little Stint** *Calidris minuta*
Non-breeding, Namibia: Andy & Gill Swash.
Adult breeding, Spain: Greg & Yvonne Dean.
Juvenile: Nic Hallam www.farmoor-birding.com.

**Curlew Sandpiper** *Calidris ferruginea*
Non-breeding, Namibia: Andy & Gill Swash.

PLATE 12

**Wood Sandpiper** *Tringa glareola*
Non-breeding, Namibia: Andy Swash.

**Terek Sandpiper** *Xenus cinereus*
*Non-breeding*: Nigel McCall.

**Common Sandpiper** *Actitis hypoleucos*
Non-breeding, South Africa: Andy & Gill Swash.

PLATE 13

**Antarctic Skua** *Catharacta antarctica*
Adult on ground: Roger & Liz Charlwood.
In flight: Allan White

**South Polar Skua** *Catharacta maccormicki*
In flight: David & Nancy Massie.

**Lesser Black-backed (Baltic) Gull**
*Larus fuscus fuscus*
Adult, Oman: Annika Forsten.
Adult in flight, Oman: Dick Newell.
1st winter, Oman: Dick Newell.
1st winter in flight, Oman: Annika Forsten.

PLATE 14

**Gull-billed Tern** *Geochelidon nilotica*
Non breeding, Uganda: Greg & Yvonne Dean.
Non-breeding in flight, Sri Lanka:
Gehan de Silva.

**Lesser Crested Tern** *Sterna bengalensis*
Breeding: David & Nancy Massie.
Non-breeding: Aurélien Audevard.
In flight: Stefano Fantini.

**Crested Tern** *Sterna bergii*
Adult and juvenile: Nigel McCall.
In flight, South Africa: Andy & Gill Swash.

PLATE 15

**Black-naped Tern** *Sterna sumatrana*
Adult perched & in flight, Australia:
Andy & Gill Swash.

**Common Tern** *Sterna hirundo*
Adult non-breeding perched & in flight, South Africa: Andy & Gill Swash.

**Roseate Tern** *Sterna dougallii*
Adult, juvenile & chick, Aride: John Bowler.
In flight: Pat Lynch.

PLATE 16

**Fairy Tern** *Gygis alba*
Adult & chick, Aride: John Bowler.

**White-winged Tern** *Chlidonias leucopterus*
Various plumages in flight, South Africa:
Andy & Gill Swash.

**Saunders' Tern** *Sternula saundersi*
Non-breeding: David & Nancy Massie.
In flight: Jon Hornbuckle.

PLATE 17

**Bridled Tern** *Onychoprion anaethetus*
Adult & juvenile, Aride: John Bowler.
In flight: John H Boyd III.

**Sooty Tern** *Onychoprion fuscata*
Adult & juvenile, Aride: John Bowler.
Adult in flight, Australia: Andy & Gill Swash.

PLATE 18

**Lesser Noddy** *Anous tenuirostris*
Adult & chick, Aride: John Bowler.

**Brown Noddy** *Anous stolidus*
Adult & chick, Aride: John Bowler.

PLATE 19

**Seychelles Kestrel** *Falco araea*
Male: Justin Gerlach.
Inset: Gerard Rocamora.

**European Cuckoo** *Cuculus canorus*:
Adult, France: Gordon Langsbury.

**European Roller** *Coracias garrulus*
Adult, South Africa: Greg & Yvonne Dean.

PLATE 20

**Seychelles Black Parrot** *Coracopsis (nigra) barklyi*
All photographs, Praslin: John Bowler.

**Seychelles Blue Pigeon** *Alectroenas pulcherrima*
Adult & chick, Aride: John Bowler.

PLATE 21

**Seychelles Turtle Dove**
*Streptopelia picturata rostrata*
Aride: John Bowler.

**Madagascar Turtle Dove**
*Streptopelia picturata picturata*
Aride: John Bowler.

**Barred Ground Dove** *Geopelia striata*
Bird: John Bowler.

PLATE 22

**Seychelles Scops-owl** *Otus insularis*
Mahé: Gerard Rocamora.

**Barn Owl** *Tyto alba*
Aride: John Bowler.

PLATE 23

**Seychelles Cave Swiftlet** *Aerodramus elaphrus*
Perched: Gerard Rocamora.
In flight, Mahé: John Bowler.

**Barn Swallow** *Hirundo rustica*
Adult perched, UK: Andy & Gill Swash.
In flight: Gordon Langsbury.

**Tree Pipit** *Anthus trivialis*
Adult, Israel: Andy & Gill Swash.

PLATE 24

**Seychelles Paradise Flycatcher** *Terpsiphone corvina* Male and female, La Digue: John Bowler.

**Seychelles Bulbul** *Hypsipetes crassirostris*
All photographs, Praslin: John Bowler.

PLATE 25

**Seychelles Warbler** *Acrocephalus sechellensis*
Aride: John Bowler.

**Seychelles Magpie-robin** *Copsychus sechellarum*
Aride: John Bowler.

PLATE 26

**Seychelles Fody** *Foudia sechellarum*
Cousin: John Bowler.

**Madagascar Fody** *Foudia madagascariensis*
Male: Nigel McCall.
Female, Aride: John Bowler.

PLATE 27

**Seychelles White-eye** *Zosterops modesta*
Head: Gerard Rocamora.
Whole bird: David & Nancy Massie.

**Seychelles Sunbird** *Nectarinia dussumieri*
All photographs: Aride: John Bowler.

PLATE 28

**Common Waxbill** *Estrilda astrild*
Adult: Brazil: Andy & Gill Swash.

**Common Myna** *Acridotheres tristis*
*Adult*: Nigel McCall.

**Feral Pigeon** *Columba livia*
South Africa: Andy & Gill Swash.

## MAMMALS

PLATE 29

**Seychelles Fruit Bat** *Pteropus seychellensis*
All photographs, Aride: John Bowler.

**Sheath-tailed Bat** *Coleura seychellensis*
Roosting in cave: Justin Gerlach.

PLATE 30

**Ship Rat** *Rattus rattus*
Adult: Mark Lucas.

**Norway Rat** *Rattus norvegicus*
Adult: Colin Carver (Windrush).

**House Mouse** *Mus domesticus*
Richard Revels (Windrush).

PLATE 31

**Common Tenrec** *Tenrec ecaudatus*
Adult:, Réunion: John Bowler.

**European Rabbit** *Oryctolagus cuniculus*:
UK: Andy Swash.

**Black-naped Hare** *Lepus nigricollis*
Gehan de Silva.

PLATE 32

**Sperm Whale** *Physeter macrocephalus*
Surfacing: Gianni Pavan.
Fluke: Trevor Codlin.

**Short-finned Pilot Whale** *Globicephala macrorynchus*
Dylan Walker.

PLATE 33

**Spinner Dolphin** *Stenella longirostris*
Dagmar Fertl.

**Bottlenose Dolphin** *Tursiops truncatus*
Lori Mazzuca.

**Risso's Dolphin** *Grampus griseus*
Graeme Cresswell.

**REPTILES**

P.96: **Seychelles Skink** *Mabuya sechellensis*
Aride: John Bowler.

PLATE 34

**Aldabra Giant Land Tortoise** *Dipsochelys dussumieri*
Silhouette: John Bowler.
**Seychelles Giant Tortoise** *Dipsochelys hololissa*
Silhouette: John Bowler.
**Arnold's Tortoise** *Dipsochelys arnoldi*
Silhouette: John Bowler.

PLATE 35

**Hawksbill Turtle** *Eretmochelys imbricata*
All photographs, Aride: John Bowler.
**Green Turtle** *Chelonia mydas*
Justin Gerlach.
**Turtle carapaces**: ILLUSTRATIONS: Rob Still.

PLATE 36

**Seychelles Black Mud Turtle**
*Pelusios subniger parietalis*
All photographs: Justin Gerlach.
**Seychelles Yellow-bellied Mud Turtle**
*Pelusios castanoides intergularis*
All photos: Justin Gerlach.

PLATE 37

**Stripeless Day Gecko** *Phelsuma astriata*
All photographs, Aride: John Bowler.
**Long Island Day Gecko** *Phelsuma longinsulae*
All photographs, Mahé: John Bowler.
**Sundberg's Day Gecko** *Phelsuma sundbergi*
All photographs, Praslin: John Bowler.

PLATE 38

**Pacific Gecko** *Gehyra mutilata*
All photographs, Mahé: John Bowler.
**Common House Gecko** *Hemidactylus frenatus*
Bird: John Bowler.
**Seychelles Sucker-tailed Gecko**
*Urocotyledon inexpectata*: Justin Gerlach.

PLATE 39

**Bronze Gecko** *Ailuronyx seychellensis*
All photographs, Aride: John Bowler.
**Dwarf Bronze Gecko** *Ailuronyx tachyscopaeus*
All photographs, Praslin: John Bowler.

PLATE 40

**Giant Bronze Gecko** *Ailuronyx trachygaster*
Praslin: John Bowler.
**Tiger Chameleon** *Chamaeleo tigris*
Mahé: John Bowler.

PLATE 41

**Seychelles Skink** *Mabuya sechellensis*
All photographs, Aride: John Bowler.
**Wright's Skink** *Mabuya wrightii*
All photographs, Aride: John Bowler.

PLATE 42

**Brauer's Skink** *Janetaescincus braueri*
Silhouette: John Bowler.
**Gardiner's Skink** *Pamelascincus gardineri*
Aride: John Bowler.

PLATE 43

**Seychelles House Snake** *Lamprophus geometricus*
All photographs: Tim Cane
**Seychelles Wolf Snake** *Lycognathophis seychellensis*
All photographs: Aride: John Bowler.
**Brahminy Blind Snake** *Ramphotyphlops braminus*
Aride: John Bowler.

**AMPHIBIANS**

PLATE 44

**Seychelles Tree Frog** *Tachynemis seychellensis*
Male: Adrian Skerrett.
Female: Justin Gerlach.
INSET: Female, Praslin: John Bowler.
**Mascarene Frog** *Ptychadena mascariensis*
La Digue: John Bowler.

PLATE 45

**Thomasset's Frog** *Nesomantis thomasseti*
Justin Gerlach.
**Seychelles Frog** *Sooglossus sechellensis*:
Justin Gerlach.
**Palm Frog** *Sooglossus pipilodryas*
Justin Gerlach.
**Gardiner's Frog** *Sooglossus gardineri*
Justin Gerlach.

PLATE 46

**Big-headed Caecilian** *Grandisonia alternans*
Justin Gerlach.
**Small-headed Caecilian** *Grandisonia brevis*
Silhouette: John Bowler.
**Flat-headed Caecilian** *Grandisonia larvata*
Silhouette: John Bowler.
**Caecilian heads**: ILLUSTRATIONS: Rob Still, based on sketches by Justin Gerlach.

**INVERTEBRATES**

PLATE 47

**African Monarch** *Danaus chrysippus*
Male upperside & underside: Gehan de Silva.
**Diadem** *Hypolimnas misippus*
Male upperside & underside, Aride: John Bowler.
Female upperside & underside: Amber Habib.
**Painted Lady** *Vanessa cardui*
Upperside & underside: Rob Still.

PLATE 48

**Evening Brown** *Melanitis leda*
Mahé: John Bowler.

**Striped Policeman** *Coeliades forestan*
Aride: John Bowler.

**Olive-haired Swift** *Borbo borbonica*
Aride: John Bowler.

**King of Seychelles** *Euploea mitra*
Museum specimen: Justin Gerlach.

PLATE 49

**Long-tailed Blue** *Lampides boeticus*
La Digue: John Bowler.

**Lang's Short-tailed Blue** *Syntarucus pirithous*
La Digue: John Bowler.

**African Grass Blue** *Zizeeria knysa*
Aride: John Bowler.

**Tiny Grass Blue** *Zizula hylax*
La Digue: John Bowler.

PLATE 50

**Convolvulus Hawk-moth** *Agrius convolvuli*
Andy & Gill Swash.

**Oleander Hawk-moth** *Daphnis nerii*
Nigel Cleere.

PP. 134–137 **Dragonflies and damselflies key**:
ILLUSTRATIONS: Rob Still, based on sketches by Bill & Chris Wain.

PLATE 51

**Black Mountain Damsel** *Allolestes maclachlanii*
Male & female, Mahé: Chris & Bill Wain.

**Blue-eyed Mountain Damsel** *Leptocnemis cyanops*
Male & female, Mahé: Chris & Bill Wain.

**Common Pond Damselfly** *Ceriagrion glabrum*
Male & female, Mahé: Chris & Bill Wain.

PLATE 52

**Marsh Bluetail** *Ischnura senegalensis*:
Male, female & fem. orange, Mahé: Chris & Bill Wain.

**Pygmy Wisp** *Agriocnemis pygmaea*
Male, female reddish & female yellow thorax, Mahé; female; Praslin: Chris & Bill Wain.

**Oceanic Fineliner** *Teinobasis alluaudi*
Male, Mahé/Silhouette: Chris & Bill Wain.

PLATE 53

**Ghost Duskdarter** *Zyxomma petiolatum*
Male, Mahé: Chris & Bill Wain.

**Seychelles Cascader** *Zygonyx luctifera*
Male, Mahé; female, Silhouette: Chris & Bill Wain.

**Phantom Flutterer** *Rhyothemis semihylina*
Male, Mahé: Chris & Bill Wain.

**Black Percher** *Diplacodes lefebrvei*
Male; & female, Mahé: Chris & Bill Wain.

PLATE 54

**Voyaging Glider** *Tramea limbata*
Chris & Bill Wain.

**Vagrant Emperor** *Anax ephippiger*
Aride: John Bowler.

**Wandering Glider** *Pantala flavescens*
John C. Abbott.

**Twister** *Tholymis tillarga*
Male, Praslin: Chris & Bill Wain.

PLATE 55

**Seychelles Duskhawker** *Gynacantha stylata*
Male, Praslin: Chris & Bill Wain.

**Amberwing Emperor** *Anax guttatus*
Male, Silhouette: Chris & Bill Wain.

**Blue Percher** *Diplacodes trivialis*
Male, Aride: John Bowler.

**Seychelles Skimmer** *Orthetrum stemmale wrightii*
Male & female, Mahé: Chris & Bill Wain.

PLATE 56

**Ant-lion** *Myrmeleon obscurus*
All photographs, Aride: John Bowler.

**Click Beetle** *Alaus scotti*
Justin Gerlach.

**Coconut Beetle** *Orcytes monoceros*
Aride: John Bowler.

PLATE 57

**American Cockroach** *Periplanetta americana*
Mahé: John Bowler.

**Yellow Wasp** *Polystes olivaceus*
La Digue: John Bowler.

**Carpenter Bee** *Mesotrichia incerta*
Aride: John Bowler.

**Potter Wasp** *Delta alluaudii*
Aride: John Bowler.

PLATE 58

**Giant Millipede** *Sechellus seychellarum*
Aride: John Bowler.

**Giant Centipede** *Scolopendra subspinipes*
Aride: John Bowler.

**Whip Scorpion** *Tarantula scaber*
Aride: John Bowler.

PLATE 59

**Palm Spider** *Nephila inaurata*
Male; female: Aride: John Bowler.

**Giant Crab Spider** *Heteropoda venatoria*
Mahé: John Bowler.

PLATE 60

**Giant Mangrove Crab** *Cardisoma carnifex*
Silhouette: John Bowler.

**Robber Crab** *Birgus latro*
Aride: John Bowler.

Plate 61

**Purple Hermit Crab** *Coenobita brevimanus*
Aride: John Bowler.

**Orange Hermit Crab** *Coenobita perlatus*
All photographs: Aride: John Bowler.

**Pale Hermit Crab** *Coenibita rugosus*
Aride: John Bowler.

Plate 62

**Horned Ghost Crab** *Ocypode ceratopthalmus*
All photographs: Aride: John Bowler.

**Chunky Ghost Crab** *Ocypode cordimana*
Aride: John Bowler.

**Fiddler Crab** *Uca* sp.
Mahé: John Bowler.

Plate 63

**Swift-footed Crab** *Grapsus tenuicrustatus*
Normal form, Aride: John Bowler.
Red form, Silhouette: John Bowler.

**Red Rock Crab** *Geograspus stormi*
Aride: John Bowler.

**Orange Land Crab** *Geograpsus crinipes*
Aride: John Bowler.

Plate 64

**White Slug** *Vaginula seychellensis*
Praslin: John Bowler.

**Grey Slug** *Vaginula voeltzkowi*
Mahé: John Bowler.

**African Land Snail** *Achatina immaculata*
Mahé: John Bowler.

**Coco de Mer Snail** *Stylodonta studeriana*
Praslin: John Bowler.

# Index of local names

This index includes the local names of all the species mentioned in the text, where this is known.

Figures in ***bold red text*** refer to the plate number on which the species is illustrated.

**Bold black** figures indicate the pages on which the main text for the species can be found.

# Index of English and scientific names

This index includes the common English and scientific names of all the species mentioned in the main text.

Figures in **bold red text** refer to the plate number on which the species is illustrated.

The **Bold black** figures indicate the page on which the main text for the species can be found.

*Italicized red* figures indicate other page(s) on which a photograph or illustration of the species can be found.

Figures in normal black text indicate other page(s) with information relating to the species.